ALSO BY ROBERTO CASATI

Holes and Other Superficialities
(with Achille Varzi)

THE SHADOW CLUB

THE SHADOW CLUB

The Greatest Mystery in the Universe—Shadows—and the Thinkers Who Unlocked Their Secrets

ROBERTO CASATI

Translated from the Italian
by Abigail Asher

Alfred A. Knopf New York 2003

THIS IS A BORZOI BOOK
PUBLISHED BY ALFRED A. KNOPF

www.aaknopf.com

Originally published in Italy as *La scoperta dell'ombra*
by Arnoldo Mondadori Editore S.p.A., Milan, in 2000.

Library of Congress Cataloging-in-Publication Data

Casati, Roberto, 1961–
[Scoperta dell'ombra. English]
The Shadow Club : the greatest mystery in the universe—
shadows—and the thinkers who unlocked their secrets /
by Roberto Casati ; translated from the Italian by Abigail Asher.
—1st American ed.
p. cm.
Includes bibliographical references and index.
ISBN 0-375-40727-8
1. Shades and shadows—History. I. Title.

QA519 .C3713 2003
535'.4—dc21 2002190850

Manufactured in the United States of America
First American Edition

CONTENTS

THE SHADOW CLUB

I In the Beginning There Was Shadow

> The Earth was without form, and void;
> And darkness was on the face of the deep.
> —Genesis 1:2

> Darest thou, monster,
> Here beside beauty
> Under the eye of great
> Phoebus to show thee?
> Come, only step forth, notwithstanding,
> For the hideous sees he not,
> As his holy eye has not
> Yet alighted on shadow.
> —Goethe, *Faust*

A Very Relaxed Beginning

The first time I watched a lunar eclipse carefully was shortly after midnight on April 4, 1996. If you ever find yourself observing such an eclipse, you should know that it's important to have a very comfortable chair. I'm no great fan of long sessions outdoors on humid country nights, and I probably would not have watched the eclipse if I hadn't had the option of doing it easily from my home in Paris. The moon was perfectly framed in the window (at the time, I lived at the top of a skyscraper), and it shone magnificently, despite pollution from the light of the *ville lumière.*

I had thought that the interesting part of an eclipse would be the shadow of the earth slipping slowly across the moon, and the chance to look into the sky and see a shaft of black negative light. This expectation was indeed fulfilled: right on schedule, the earth's shadow cone dimmed and then completely extinguished the moonshine. But at that point, in the moment of total eclipse, I had—pardon the expression—an illumination.

For the first time, I *saw* the moon for what it really was, and I wanted to put it down in words. The moon is a fairly large, shadowy rock hanging a certain distance over my head; oddly, it doesn't fall down and hit me. Naturally, I knew about the laws that kept it safely in orbit; but my eyes, unaccustomed to seeing stones hanging in the sky, didn't want to believe it. Likewise, it had escaped my notice that the moon was—as I knew perfectly well—a huge dark rock; usually the diaphanous glow of the lunar surface tricks the eye with the illusion that it's a delicate, airy lantern.

During an eclipse the moon loses its character as demigoddess: it splits itself off from the royal court of the other visible, shining celestial objects. Even those planets that are dark like the moon and glow with reflected light aren't seen as planets: our not very selective vision lumps them in together with the stars. Light offers the moon a weightlessness that renders it more acceptable—makes it seem almost normal that the moon should sail in the night like a paper lantern hung from the black ceiling of the sky.

So the shadow of the earth reveals the true nature of the moon.

Since that night I haven't stopped thinking about shadows. Up till then I had considered them to be an interesting example of a strange, not very clearly identified object, with some philosophical complexities. (Strange objects are always thought-provoking for philosophers.) Beyond this narrow theoretical interest, shadows simply registered on me with the low, negative associations that our culture has (and many other cultures have) bound them up with: that shadows are for hiding in and plotting in; that shadows are dark and worrisome. All of a sudden, as I watched the eclipse drawing to a close, shadows seemed much more important and worthy of attention. If shadows can forever change the way we think about the moon—expelling it from the Olympus of the gods to imprison it in the world of minerals, in a spectacular comedown, the opposite of the ascent through the chain of being—then shadows are valuable tools of knowledge. They don't hide: they reveal.

In other words, too much light can be harmful. And yet some people can never get enough light, as we know from one of the most famous images in Western literature, described below.

A Completely Different Beginning

We enter a corridor that leads underground. In a poorly lighted room are the subjects of the experiment. Ever since birth they have been kept in the dark about what happens outside. For even greater security they are in shackles. Looking more closely, we see that they are bound in such a way that they cannot see one another and they are all forced to look in the same direction. The only distraction—which is, however, a crucial part of the experiment of which the prisoners are, unbeknownst to them, the subject—is a strange spectacle taking place on the spot they all see. The subjects don't know that their lives are exceptional, and they have spent many years watching the scene without the option of looking away. It is explained to us that the experiment has been a success so far: the subjects believe that the show is the sum total of reality, and the actors are the only inhabitants of the real world. (At this point we wonder what the show is. We would like to understand what the prisoners have learned in all these years.) We discover that the designers of the experiment have created a clever scene. The actors are not flesh-and-blood people with their own voices and their own free will, but rather statues brought to life by an army of attendants going up and down a catwalk hidden behind a low wall. And that's not all. The prisoners don't even see the statues. The low wall over which the statues protrude is located not in front of the prisoners but behind them. A bonfire makes the statues project a shadow on the front wall, the wall that prisoners are made to face.

The prisoners have always seen only shadow theater, and this is why their minds are a thousand miles away from the real world. Of course, only a rough caricature of the universe would have been provided by a normal stage show or, even more roughly, by a show of statues or marionettes. But a shadow theater provides even less: it's the image of an image.

But what's the point of this complicated staging? We are told that the crucial phase of the experiment has yet to begin. Keeping the prisoners in shackles for all these years has served to distance them from reality. What will happen when one of these prisoners is given the opportunity to leave the bowels of the earth and discover the world of things and the sunlight that gives those things shape and color? The hypothesis is that the prisoners exiting the cave will first be blinded by the light and then, when they can see again, will marvel at the existence of a world of colorful three-dimensional things. For too long they have been deceived by the flat and colorless world of shadows.

This experiment, fortunately only imaginary, is described in the seventh book of the *Republic* by Plato (427–347 B.C.), written more than two thousand years ago, between 388 and 367 B.C. What Plato wanted to suggest with the picture of the cave is that we are like the prisoners in the experiment, even though we consider ourselves to be free and well informed and even though we haven't spent our lives in chains watching shadow theater. The prisoners see the shadows, they don't believe anything else exists, and they lose the chance to know about real objects. In turn, we who see objects (and not just shadows) are unable to consider their true hidden nature, which would mean digging under the surface of things. Our consciousness, like that of the prisoners, is locked up in an image. And what if a prisoner emerges into the light and then returns to the cave and tries to convince those inside that a better world exists? He will only be ridiculed by his former companions. His vision will be weakened when he goes back into the cave, and his companions will think that the light he praises so highly is actually damaging. According to Plato, this is what will happen to the philosopher who tries to explain the truth to other humans.

Now let's try to focus on the situation of the prisoners.

The first fact is that there was no reason to use shadows for this experiment. The statues, the marionettes, would have sufficed. To be perfectly logical, even flesh-and-blood actors would have served to make us ponder the distinction between image and reality. Why did Plato choose shadows? One answer is that shadows are a disturbing example of lesser entities: they are a *diminution* of the objects that project them. They are flat, immaterial, and without qualities, without color. Their outlines enclose an indistinct interior. But above all they are *absences, negative things.* A shadow is a lack of light. Negative things are bizarre. In the end, actors and attendants are on the payroll of the theater. Even marionettes are counted, during inventory. But shadows are not listed anywhere. These uncertain beings confuse the mind; they trouble us. As if that weren't enough, shadows have always been surrounded by suspicion and fear. They are strange entities: we don't know much about them, but we do know that they're not a happy bunch. Plato is right to use them if he wants to worry us.

There's more. For Plato it went without saying that shadows keep us at a certain remove from knowledge. They are a screen between the prisoners and the truth. And yet, despite their unhappy condition, despite the fact that the theater's paltry tricks impose a distance between them and reality—despite all this, the prisoners have some reason for optimism. They can study the shadows carefully. They can reconstruct the three-dimensional forms of the objects casting the shadows. They can grasp the mathematical beauty of the relationship between an object and its shadow. Approached in a certain way, the shadows turn out to be a magnificent tool for knowledge. This is why in the final analysis Plato's experiment is not convincing. Shadows can be used to reconstruct the world. And indeed we continually use them to understand the shape of our environment.

Just Imagine a World Without Shadows

If light is the instrument of vision, shadow must be its great antagonist. One hides in the shadows because a searcher's gaze cannot penetrate the darkness. But it's also true that vision can't do without shadow: the information carried by shadow is a fundamental aid to seeing. Evolution has recognized that the world is full of shadows and it has selected biological systems that adapt to levels of darkness. Why do many animal species have bellies paler than their backs? Because light comes mostly from above: the pale spot on the belly counteracts the inevitable shadow. Thus the animal hardly stands out, and he is less easy to see. Here evolution was betting on the fact that the visual system of the predator was hunting for shadows.

Things could have gone differently. Permit me to develop this hypothesis, though I'll need to rewrite the laws of physics. Let's say that all the objects in the universe were dimly phosphorescent. The slight bit of light they gave off would make only weak shadows. But these shadows would fall on phosphorescent surfaces and be canceled out. In such a universe evolution would have had no reason to create eyes that notice shadow, as was the case in our world. Our vision is so bewitched by chiaroscuro that if we were to find ourselves suddenly in a shadowless world, everything would seem insubstantial and without depth.

Here's another example of a world without visible shadows. If our eyes were fitted with lamps, we would never see the shadows they cast: the shadow would always be hidden by the object making the shadow. The sun's view of the world is immaculate because the sun "has not yet alighted on shadow," as Goethe wrote in *Faust.*

Immaculate it may be; but it certainly is flat. Rooms and faces photographed with a flash are flat. In images where we have erased the shadows, objects are flat and they lack context.

The opposite is also true: erasing the light eliminates relief. In what may have been the first experiment ever conceived of in cognitive science, Galileo Galilei (1564–1642) suggested blackening the unshadowed parts of a statue in order to eliminate the chiaroscuro. His hypothesis was that the statue would appear to be without relief—flat. Galileo wanted to show the superiority of painting over sculpture. In modern terms, the experiment shows that it's not enough to have good volume in order to look like an object with good volume: you also have to make the right impression. Putting on makeup does the same thing: augmenting the darkness of a part that's slightly shadowed gives your face the appearance of greater depth.

Cast shadows are removed from this painting by Jean-Baptiste-Siméon Chardin, Copper Urn. *As a result, the objects "float." Then the contrast is eliminated from the picture, and the picture lightened. As a result, the objects lose their solidity. (Graphic manipulation by W. Criscuoli.)*

But what can be said about the world that so frightened Plato, a world in which perception depends on shadows exclusively? We know part of the answer. In 1953 the psychologists Hans Wallach and D. N. O'Connell actually carried out a variation on the cave experiment. First they took a piece of stiff wire and bent it irregularly. Then they hid it behind a screen and had a lamp project a shadow of the wire. All the observers saw was the shadow of the bent wire on the screen. If the wire stayed still, observers would see a stringy black mark like a hand-sketched line. But the wire was actually mounted on a turntable. As soon as the turntable began to move, the projection of the shadow on the screen changed. Just as in Plato's cave, all the observers had to work with were the dancing shadows, two-dimensional figures playing tag. But they did not in fact perceive a deformed black mark stretching and shrinking: they had the distinct impression of a three-dimensional wire spinning in space. The brain works feverishly to extract solid structures from a changing reality of shifting perspectives. One might say these observers were different from the prisoners in the cave; Wallach and O'Connell's observers were not kept in chains and obligated to look *only* at shadows from infancy. Their brains could have learned to see three-dimensional things, while those of Plato's prisoners might have been affected by their strange upbringing in the cave. But it's also true that for countless generations our ancestors' brains were immersed in a world of rigid three-dimensional objects, and we inherited a propensity to see solid things. This propensity kicks in on the slightest pretext and makes us see objects even where there is nothing but shadows. In a world made entirely of shadows, our brains might manage better than Plato suspected.

Certainly, we now know more about perception than was known in ancient Greece (although the experiment by Wallach and O'Connell requires equipment so simple that it could have been done in the fifth century B.C.: just replace the turntable with a turning lathe). Are we being unfair to Plato?

About This Book

Or is it perhaps Plato who is unfair to shadows? The fact that we use shadows involuntarily and automatically in order to perceive things in space is only one aspect of their link with knowledge. Conscious use has been made of shadows ever since ancient times, notwithstanding the fact that shadows were feared and that no one quite knew what they were. The history of science is woven together with the fabric of shadow. It's interesting to test the density of that fabric.

To do so we must work on two fronts. On one hand we must understand why shadows are considered perilous for the mind. If we tried to describe their true nature we would reach a dead end. If shadow is absence, a thing that does not exist, then it doesn't exist and that's the end of it. But if that's the case, how can we be talking about it? Maybe it's something more than an absence, or maybe it's just an illusion? Shadows certainly are mysterious. On the other hand, despite their being so precarious, and despite their being so mysterious, shadows are a valuable aid to knowledge. How can we reconcile these two perspectives?

This book is an attempt to show that shadows are not bad fellows at all, even though they don't seem very trustworthy at first glance.

When I compiled a list of shadow discoveries, I was surprised. The list is quite long; I have included it as an appendix to this book. I've chosen not to talk about every single discovery, but to tell a story that traces the uninterrupted dialogue, many millennia long, between us terrestrials and the sky. I was surprised to see how many members are enrolled in the Shadow Club—the company of famous and less-known characters, from Eratosthenes to Galileo, from Arab astronomers to modern mathematicians, from Greek painters to Leonardo—who have made shadow the friend of knowledge. I was surprised to see how a confused concept with no great ambitions could be exploited so astutely.

I was surprised also whenever I spoke about shadows with friends in a café, or with my students and my teachers, or even with strangers I met on the train, discovering something new every time. Shadows are marvels for the mind. You think you can say everything about them in just a few lines, but if you scrutinize them carefully, looking deep into their heart of darkness, they turn out to be infinitely complex.

Part One THE HEART OF SHADOW

(Curtain rises)

Plato and His Shadow

The foot of the Acropolis. Plato is heading toward the sea. Skia, his shadow, is just barely visible in the midday sun. The cicadas are screaming, maddened by the heat.

SKIA: How tiring! Can't we stop and rest?

PLATO: What are you saying? Rest! I'm the one who's walking. Your ridiculous prancing is nothing but an imitation of my stride.

SKIA: I'm not walking, but you keep stepping on my toes!

PLATO: So what? You're nothing more than a shadow. You're not made of flesh and blood; you can't feel pain. I don't even know why I'm talking to you—maybe with this heat I can't see straight.

SKIA: Well, you wouldn't turn your nose up at the freshness that my sisters offer you. We could stop for a moment in the shadow of that grotto down there.

PLATO: Never, never! I'd rather melt in the sun. I make an enormous effort to lift humanity out of the darkness. This is not the moment to turn our backs on the light.

SKIA: It's clear as day that you don't like me. But we still have a long way to go together.

PLATO: I would happily go without you.

SKIA: Why are you so set against shadows? What did they ever do to you?

PLATO: They're too intrusive, that's what. They're distracting. They're dark. They frighten children. They're hard to understand. They create all kinds of problems.

SKIA: Can you give me an example?

PLATO: Read this and you'll see.

II Ancient and Modern Shadows

> I know so many people who are more afraid
> of the shadow than what casts the shadow.
> —Abraham B. Yehoshua, *Mr. Mani*

Shadow and light are linked in the history of technology. Electrification erased the vast shadowy zones that made cities unsafe. It is our inheritance from the nineteenth century, which saw a radical improvement in lighting. In the space of sixty years, from 1820 to 1880, numerous kinds of lamps were invented, all of them easily powered and relatively economical. Until the end of the previous century lights were fueled primarily with whale oil, olive oil, and wax. At the beginning of the 1800s, lamps using natural gas and coal gas became popular. Streets were lighted with gas lamps in the major European and American cities from the very beginning of the century. The digging of the first petroleum well in 1859 made another combustible source available. Paralleling these advances was the discovery that electricity could be a source of light. At first a bolt of lightning was the model, in lamps that had an electric arc leaping across the gap between two carbon electrodes; this system was perfected by the Russian engineer Pavel Yablochkov around 1875. The next step was the filament bulb, an invention that was in the air generally but which is now linked to Thomas Edison (1847–1931). Edison tried a thousand times with filaments of different materials (sealed in a vacuum tube) that were made incandescent by the passage of an electric current. On October 21, 1878, he decided to try a filament of burning carbon, which gave a stable glow for a couple of days. It was the solution he had been looking for. After several months of testing, in May 1880 he launched the era of electrical lighting, which would in just a few years replace gas lighting entirely.

Electricity is easily transported and its fire risk is low—this was a remarkable improvement over torches, candles, and the saucers full of combustibles that had been used in previous centuries. But the progress was not limited to these technical aspects only. The quality of the light was completely different. The same basic phenomenon, incandescence, is produced by both candle flame and lightbulb. The matter loses part of its energy when heated, releasing a flow of photons. At higher temperatures the balance of intensity of the light emitted tilts closer to blue, and light in the room looks less reddish; the weaker candle has a redder light. It should be said that the gas lamps used since 1820 already had an important technological improvement: the principal source of illumination was not the flame but a chip of fireproof material that was heated by the flame. Thus a static component gave off the light, as in a modern lightbulb, where the filament is heated by a current of electricity passing through it, a current that the filament resists. This allows for higher temperatures and a greater release of photons.

So the newer light sources were more luminous, and they also had another property: they were *stable.* They no longer depended on a flame exposed to air currents, and they didn't flicker. This has extraordinary consequences for our story. As if by magic, shadows too stopped flickering along the streets and within the houses.

The nineteenth century didn't just vanquish shadows; it created *new* ones. They were the frozen shadows produced by a fragment of material heated to incandescence. These were new shadows: static shadows had never existed in nature, nor were they ever before produced by man.

Until just a few generations ago, shadows were always moving. No shadow was ever really still. Candles and hearths project shaky, agitated shadows on the walls of a room. Outdoors, all you have to do is trace an outline and then come back to it a few minutes later to see the movement of apparently static shadows cast by bodies in sunlight. Sundials work because of the movement of shadows. Artists have always had great difficulty painting landscapes or buildings lit by the sun. After an hour passes, the distribution of shadows in the landscape has changed so much that the view is unrecognizable. Partly for this reason, painting classes study the *theory* of shadows, which frees objects from the constant mutation of natural chiaroscuro. The pioneers of photography faced a similar problem. The photographic technique invented by Joseph-Nicéphore Niepce (1765–1833) uses a bituminous substance that becomes insoluble when exposed to light. Unfortunately, eight hours of exposure is required, and in that time shadows move quite far around the subjects. If you want to freeze a shadow, you can make do with *shadowgraphs*, as did William Henry Fox Talbot (1800–1877): you immerse paper in sea salt and then in a silver nitrate solution, you artistically lay leaves, cutouts, and lace directly onto the surface, and you allow the sun to darken the bare parts while leaving the covered parts white. But then the only objects that you can capture in the image are just as flat and devoid of nuance as the shadows that represent them.

Modern shadows—stuck to walls, jammed between houses—are like a new species that has populated the earth by colonizing the empire of the night. And yet, even though they have sneaked in everywhere, they haven't managed to unseat the living shadows that have accompanied our species for millennia. Nowadays we don't time an appointment for the hour when our shadows are twice as long as our bodies, but we still sense the ineluctable force of shadows lengthening and shrinking in the course of the day. As every second passes, countless objects—pine needles, rocks, insects, people—cast shadows. As every second passes, these shadows shift imperceptibly but inexorably.

A typical shadowgraph image

The ancient shadow is always, slowly, moving.

Tokyo and New York

The modern shadow is produced by a quick and violent hand. In Hiroshima the heat wave from the atomic explosion dissolved the surface of the buildings. Half a kilometer from the blast a man waiting for a bank to open protected the building wall for an instant. That instant was long enough to mark a difference between the area directly exposed to the heat and the area blocked by the body of this bystander. This difference registered as a shadow on the wall. The final act of the man's life was to leave a shadow that survived him. But Japan deserves to be remembered for better shadows. The most famous ode to ancient shadows comes from the pen of a Japanese writer, Junichiro Tanizaki (1886–1965), who contrasted the warm, intimate shadows of the Japanese house with the cold illumination of the West—the illumination that produces modern, standardized, and uselessly sharp shadows. Tanizaki laments the disappearance of ancient Japanese civilization—for him, the abandonment of shadows is an unmistakable symptom of decadence.

The facts show, however, that Japan is far more shadow-phobic than Tanizaki suspected, and that the fear of shadow was to create some difficulties in the East. Such is the view of Lester Thurow, an MIT economist, regarding the factors that will keep Japan mired in the economic crisis of the end of the millennium. "Japan might have brought about a housing boom to restart its economy if it had been willing to change some of its obsolete regulations and tax laws. . . . The current requirement that people building high-rise residential towers must negotiate with and compensate those that lie within their shadows essentially stops Japan from solving its urban housing problems as other countries have." Tokyo is not an isolated case. Antishadow regulations are in force everywhere, and they have affected the development of cities. The most spectacular case was perhaps the skyline of New York between the two world wars. In photographs from around 1940, New York looks like a Mayan city.

New York or Chichén Itzá?

The tips of the great towers narrowed in huge steps; the skyscrapers were topped with pyramids. Why? The skyscrapers and apartment blocks cast enormous shadows from their tips down onto the streets, but the opposite is also true: the skyline of Manhattan is a drawing projected from below, from the bottom of the streets, starting with the shapes that the buildings' shadows *must not* cast. Shadow has determined the profile of the city.

The problem first arose at the beginning of the twentieth century, and it exploded between 1912 and 1915 with the construction of the Equitable Building (at 120 Broadway) by the architect Ernest R. Graham. It's a strange skyscraper of thirty-six stories in neo-Renaissance style, 166 meters high, that occupies an entire Manhattan block. From above it looks nothing like a skyscraper—the plan forms an H—and that's the problem. The Equitable Building is not particularly tall for New York, but it's very wide. If all the builders in New York had imitated it, the citizens of Manhattan would have ended up walking along the bottom of gloomy canyons. The owners of nearby buildings and land were worried by the plans for the Equitable, so worried that the projected sixty-two original stories were reduced to forty (and later reduced again, to thirty-six, to optimize the use of elevators). But when the building went up, the neighbors fled. The city took a big financial loss, and it decided to react with a strict regulatory plan that required the tops of buildings to taper as they rose. Skyscrapers could continue to grow indefinitely once the shrinking floors filled only one quarter of the structure's footprint on the ground. The extraordinary shape of Manhattan between the wars derived from an absolutely slavish application of zoning regulations in order to obtain the most usable space possible. But the advantages were remarkable for both the developers (stable land prices) and the city (reliable fiscal yield). Skyscrapers topped with points or unlikely pyramids and ziggurats were made obsolete in 1958 by a mighty move on the part of the architect Ludwig Mies van der Rohe (1886–1969). Offering the city a plaza at the bottom of the canyon, he was able to send his glass and metal box, the Seagram Building, straight up without a single setback from the base to the roof line. Each morning the smooth black parallelepiped throws its shadow with impunity onto its *own* plaza.

In spite of the regulatory plans, New York is still a city of shadows. It's impossible to shoot a satisfactory aerial picture of Manhattan. The buildings are so tall and so densely grouped that even at the best moment, midday on the summer solstice, grim shadows ruin the photograph. New York at night is a city of light, but history has forced it to hide in darkness. During the Second World War, Manhattan had to be blacked out because the German submarines off the American coast could see—against the city's luminous halo—the aid convoys heading for the United Kingdom.

Never underestimate the power of shadows.

Shadow Theater

To closely examine the mechanism of Plato's cave, I shut myself inside a large dark space.

The working lights are off. I'm standing still in the wings. I hear the actors' footsteps and I glimpse their figures. The action begins, a spotlight is turned on, and an actor makes a move that seems incomprehensible to me at first. I lean out, and I can see his gigantic shadow fluctuating across the scrim that separates him from the audience. All at once, the actor's gestures are clear. Before, I hadn't understood what he was doing because his actions didn't have a direct effect on anything touching his body. The actor is acting at a distance, like a conductor leading an orchestra of images. He is moving his shadow.

I have asked a shadow theater troupe to let me watch their show from the viewpoint of the actor. I wanted a close-up view of how they design and manage shadows. (The show is a version of *Alice in Wonderland,* a subject that lends itself to shadow performance: in the story Alice changes size, becoming tiny and then gigantic in just a few moments. This is easy to do with shadows: you just have to go farther from or closer to the light.)

The origin of shadow plays is lost in the mists of time; but there were still traces of the ancient shows in recent times, when candlelight was the only source of illumination, and a crowd of shadows surrounded any table full of diners. There is documentation of shadow plays existing in India and in China around A.D. 1000; such plays spread and flourished in Java (where the term *wajang,* "shadow," is also the term for "theater," and actors are called *wajang wong*—shadow men). The theaters of Bali, Sumatra, and Borneo must have derived from that of Java. The shadow is made not by an actor but by a translucent colored figure cut out of leather. In the Middle East such theater is of Arab origin; perhaps it spread from Egypt to Turkey, Greece, and northwestern Africa. In Turkey the chief character, Karagoz ("Black Eyes") is the hero of salacious adventures based on the shadow of his embarrassing phallus, which is often mistaken for the equivocal shadow of some other object.

Behind the comedy, though, are hints of a deeper meaning. The subtle message of this theater is that man is just a shadow in the hands of the creator, an idea that is the basis of the Sufi school of Islam.

The earliest Western documentation of Eastern-derived shadow theater dates from the end of the 1600s. In France in 1767 people began to speak of "Chinese shadows." The popular taste for this genre was confirmed by the success of the Séraphin dynasty—Dominique Séraphin and his descendants—shadow performers from 1784 to 1858.

From a cognitive point of view the success of shadow theater can be explained as a response to a *need for moving images.* But in the West the theater was transformed, becoming artificial and mechanical. Whereas in the East the person controlling the image might be a priest and the audience could see his gestures as well as the sticks used to manipulate the silhouette, in the West the actor became a technician and the artifice allowing him to hide his work grew ever more sophisticated: the demands of the illusion trumped the demands of the drama. Machinery ended up destroying shadow theater: the Parisian theater Le Chat Noir closed in the very year that the Lumière brothers projected their first film. Shadows went modern and became reproducible.

Even More Modern Shadows

The contemporary rebirth of shadow theater is due to technological improvements that have changed the nature of performance. Until after the Second World War, the light source used for projection could be shrunk to make a pinpoint light only by sliding a perforated screen in front of it, but this cost it some of its brightness; without the screen, the light source was broader and thus it made unfocused shadows.

To solve this problem and to avoid a fuzzy shadow, traditional shadow theaters leaned the silhouettes right onto the screen. For example, in Java the traditional theater figures are colored semitransparent silhouettes of stretched animal skin. Transparency throws the image onto the screen. (We might ask: are we seeing colored shadows, then, or colored lighting? This is the border between the concept of shadow and the concept of light.) But the disadvantage of this device is that perspective effects are impossible. If you want to show Alice growing, and you pull the silhouette away from the screen, you get a less distinct shadow.

The new world of contemporary shadow theater is made possible by sources that emit light from a tiny area, such as the new low-tension

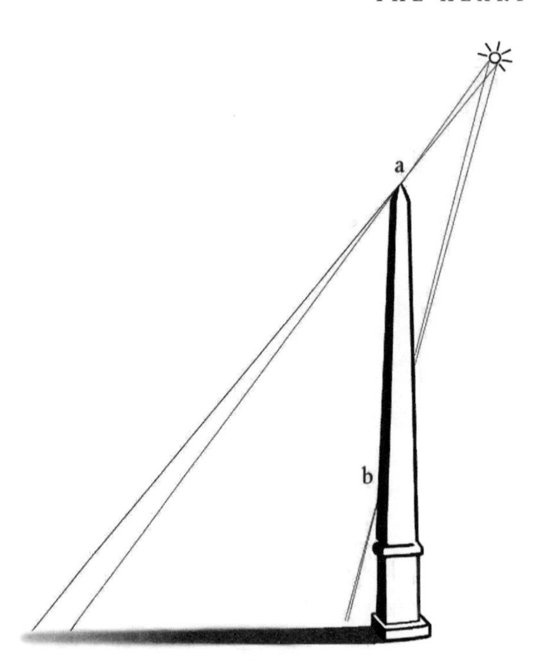

Any light (including the sun) not from a pinpoint source always produces an unfocused shadow. Point a, *which is farther from the earth, suffers more from the lack of focus than does point* b; *the shadows from point* a, *which are created by the outer edges of the light, end up being farther apart than the shadows from point* b.

beams. The shadow obtained with a nearly pinpoint light is always well defined, whatever the puppet's distance from the light or from the screen. This has two consequences in performance. First of all, the actor is free; he's no longer glued to the projection scrim. And, most of all, the shadows get detached from the bodies. Shadow actors make an important distinction between *floor shadow*—such as that projected by the sun or by a lamp—and *vertical shadow*—which appears on the screen from a light at floor level. Fabrizio Montecchi, a shadow theater director, explains that his actors begin their apprenticeships working with floor shadows. "The floor shadow follows you, it's attached to your feet. The floor shadow is *your* shadow." Then they move on to shadows unattached to their bodies, to vertical shadows. If the light source is at floor level, the floor shadow is not visible. The shadow projected onto the scrim becomes an autonomous object, a character in its own right, and it's no longer a stand-in for the original.

Thus, after much study, the theatrical shadow succeeds in faithfully representing itself. Indeed, ethnographic tradition and psychology both focus on shadows that have a life of their own.

III *Shadow Merchants*

Man is the dream of a shadow.
—Pindar

Dangerous Shadows

If the shadow of an Untouchable brushes over the body of a Brahmin, the Brahmin must purify himself.

Stories, taboos, and rituals about shadow are found in every era and in very diverse cultures. They bring us into an area of complex images born of the slow sedimentation of social forms and narrative traditions. According to some ethnologists and anthropologists, shadows would reveal how Western thought differs from non-Western thought. In Western culture, beliefs about shadow are sooner or later framed in a minitheory, and though adults may have a concept of shadow that is sometimes imprecise, still the way they think about shadow boils down to an understanding of the relationship among light source, screen, and projection wall. But certain non-Western cultures seem to prefer a conception of shadow as an object animated by a life of its own; they attribute various powers to shadow and they consider shadow to be an image of the soul. In other words, alongside the Western kind of shadow there is also a "primitive" kind of shadow. In a famous chapter of his *Golden Bough*—one of the basic texts of modern anthropology—James Frazer (1854–1941) listed some beliefs likening the properties of shadow to those of the soul, or at least to those of "a living part of man and animal." More recent ethnography brings us many examples that seem to point in the same direction. Some believe that you can act upon other people by acting on their shadows; by the same token, someone else's shadow can act upon you, and so you have to keep alert. The details of these theories are interesting. You must, for example, look out for your own shadow; a blow to your shadow can make you sick (so they believe on the Indonesian island of Wetar); you mustn't let your shadow fall across certain bad-luck boulders and then be eaten up by them (Banks Islands of Vanuatu); you mustn't let your shadow slip into an open coffin or grave (so they say in China, where gravediggers used to tie their shadows to their waists with a string for extra safety). Shadows can act as well as be acted upon, and they can be quite sneaky because they move silently. In Australia they say that while you lie sleeping unaware, someone's shadow (beware of your mother-in-law, in particular) can sweep over you and make you very sick.

The legend of Tukaitawa the warrior, told by the Mangaians of the Cook Islands, is the ultimate warning. His power grew and shrank with the length of his shadow, so all his enemy had to do was wait until midday, when shadows are shortest, to challenge and kill him.

In these stories, shadows behave like a vital part of the body (such as the heart), and so they must be protected. Other narratives make shadows into a psychic phenomenon. Let's look at legends from sub-Saharan Africa. The Yoruba maintain that the soul is represented by *òjìji,* the shadow, and that one can harm another person by "working" on his shadow. Among the Ewe it is thought that the various elements comprising the soul are put together at birth. These elements are *luvo,* the soul, *gbogbo,* the spirit, and *vovoli,* the shadow. And indeed the word *vovoli* can be used to speak of the soul, and it is interchangeable with *luvo,* because a shadow is what makes a person recognizable: a shadow is the visible form of the soul.

A sophisticated theory about shadows is found among the Dogon. For them, a shadow represents the not intelligent soul, and it expresses the soul's qualities in the play of light and dark. This soul, which can also be seen reflected in water, is considered an opposite-sex twin of its owner. It is not a vital element, because it disappears only long after death, when the body dissolves into dust and ceases to cast a shadow.

Biya, the shadow of the Songhay, travels during sleep (which explains the mechanism of dreams) and can be attacked, stolen, and devoured by witchcraft. One of the spirits of the Alladians and the Ebrié is designated by the name also used for a cast shadow. The Luba consider that a person is made up of three genuine shadows (*umvwe*). The first one, a solar

Peter Schlemihl letting his shadow be cut off. Etching: George Cruikshank, 1823

shadow, is the model for the other two; with a magical operation the soul can be shut up inside the shadow, at least temporarily, thereby rendering the body invulnerable. The second shadow is the shape of the body, and it follows the stages of the body's life: in youth it's just a sketch; it is complete only in adulthood (and if the shadow is cut, premature death will follow); and it disappears completely only when a dead body is dismembered. The third shadow is the soul itself, and after death this remains in circulation for a while. The Zulu people fear that a shortening of the shadow is an omen of death, as is proved by the puny size of a cadaver's shadow.

Western thought, too, has of course recognized the animistic role—if not the magic power—of shadow, and the vestiges survive in literary tradition. Peter Schlemihl, the unlucky protagonist of the 1813 story by Adelbert von Chamisso, sells his shadow to the devil but then finds he cannot continue with his normal life. And really, who would trust someone who doesn't cast a shadow; who wouldn't at least consider him a freak? (The same goes for someone who doesn't have a reflection in the mirror. The message is the same: it's important to *keep up appearances.*)

In James Barrie's *Peter Pan* (1904), the title character makes an escape during which he loses his shadow: it gets caught in a window, and Mrs. Darling rolls it up and puts it away in a drawer. The heroine of the opera *The Woman Without a Shadow* (1919), by Hugo von Hofmannsthal and Richard Strauss, is the daughter of Keikobad, king of the spirits, and she is sterile; her coveted shadow, which she obtains after many vicissitudes, is an indication of her newfound fertility. The saga of Walpurgisnacht immortalized in Goethe's *Faust* is linked to the shadow trick of the Brocken in the Harz Mountains northeast of Göttingen; this is a rare optic phenomenon that, at sunset, projects the shadow of any passersby onto the surface of low clouds: the shadow is enlarged and surrounded by a colored halo.

Even animals have their own shadow stories. In ancient Greece a dog could fall from a roof if his shadow in the courtyard below was stepped on by a passing hyena. An *umbrous* horse that takes *umbrage* is an animal scared of its own shadow. Alexander the Great solved that problem by placing Bucephalus, his rebellious horse, in a position looking toward the sun so that he didn't see his own troubling shadow.

Shadows Explained

A rational spirit will find a simple explanation for even the weirdest ideas about shadows. Dead people don't cast shadows? Well, obviously not, because the body is flat against the ground. And those who are ill, who likewise lie down, also cast a limited or weaker shadow. It's a short leap from noting this fact to believing that shadows can be used as a diagnostic tool. And isn't this leap characteristic of magical thinking, which easily confuses cause with effect? It's not that your shadow is short because you're lying down, but rather you are laid low by sickness *because* your shadow is short and weak.

But behind this facile explanation lies a more complex problem. We should note right away that it's not clear what can be gained from ethnographic reports. Do the people who talk to ethnographers *really believe* that shadows can act and suffer like a real body part? Some of the difficulty comes from the very nature of ethnographic research. Anthropologists have their informers who tell them about folk beliefs; but some of this information comes through as jokes and wordplay. To discover whether the informers truly believe the things they say, one would have to trap them—for example, by seeking contradictions in the stories they tell.

And there's no shortage of contradictions. If Tukaitawa the warrior is weak at midday when his shadow is shortest, isn't it also true that it's the shortest moment for his challenger's shadow? So how can the challenger be the stronger of the two at that time? Or do the Mangaians perhaps not understand simple geometry? Again, if having a shadow is equivalent to having a soul, does that mean that tables and chairs also have souls? So perhaps non-Western thought isn't troubled by contradiction—perhaps it even embraces contradiction?

Here is a cautious interpretation of the ethnographic material available to us. Yes, the shadows of people and animals are accorded more properties than would be attributed to them on the basis of optics and perspective alone. For example, a shadow not only depends on the person who casts it, but it also takes some of that person's physical characteristics such as shape and size. Yes, a shadow is practically considered to be like a thing. And yes, as an almost autonomous thing, a shadow becomes an interesting model, rich with consequences, for representing the vital center or soul of a person. But this model is of limited use. That is to say, the richness of the model is never exploited adequately, and indeed it's sharply circumscribed: in practice, people are very careful not to take their thoughts to their logical conclusions. For this reason it's never clear *how* shadows exercise their influence on people and things, nor why chairs and tables don't have souls even though they do have shadows. Mythic creations (for example, the shadow as soul) are *segregated:* the beliefs and practices built up around them don't disturb the rest of the beliefs and daily practices that must in any case function well enough to allow people to continue avoiding obstacles, building tools, and so on. This segregation heads off contradictions, and thus there is no reason to declare that the people interviewed by anthropologists have contradictory ideas.

At this point one could accuse the informers of systematic bad faith, or one could say that the shadow stories are only myths, fables, or entertaining tales. Some ethnographers have clearly indicated that beliefs about shadow are obscured in a murky cloud as soon as one moves beyond the simple classifications of souls and vital elements. On this topic, the ethnographer Jean-Pierre Olivier de Sardan writes about the Songhay that the role of shadow is evident only when the shadow is absent: "Without his shadow a man becomes gravely ill and falls into lethargy; he dies as soon as the shadow is consumed by a sorcerer. There are no *explanations* of the shadow. The magical operations are more a collection of operative recipes than they are an application of some basic science." This is exactly what is to be expected of shadows if they are considered as near-objects that don't obey precise rules. The Kokoto recognize two kinds of shadow, but one of them is invisible to most people and appears only to those who practice witchcraft. This is a belief that shows the limits of the idea of the soul as a shadow: the visible shadow isn't enough to suggest a soul—an invisible shadow has to be added to it. The Alladians and the Ebrié who were questioned explicitly by the ethnographer Marc Augé couldn't say anything precise about the origin and the destiny of the soul-shadow. After having reported that according to the Lugbara (of Uganda) witches don't have shadows and that they can harm other people by treading on their shadows (*endrilendri*), the ethnographer John Middleton hurried to observe that the Lugbara don't pay much attention to the concept of shadow.

Truly Unbelievable Things

These bad-faith mechanisms were well known to the cunning shadow merchants who, even as late as the end of the nineteenth century, reputedly still circulated in the Romanian population of Transylvania. Frazer reports that in ancient times a live person was sealed into the walls or crushed beneath the foundation stone of a new building to guarantee its stability or to ensure a ghost that would haunt the building and scare away thieves. In more recent times this practice was replaced with a somewhat less cruel ritual—burying the shadow or the length of the shadow of a passerby, who would then be sure to die within forty days (or in any case before the year was out). The shadow merchant measured other people's shadows and made them available to the architect; whenever he didn't have a consenting seller, he might cynically offer his own shadow for sale—thereby showing that he didn't completely believe in what he was doing.

Strange beliefs about shadows can also be explained another way: they are not real beliefs because they are literally unbelievable. They are all just *stories;* they are filler used to decorate other tales. Thus there is no "primitive thought" that believes shadows to be magical beings. Ethnographic studies about shadows report not beliefs but narrative traditions; they do not reflect any alleged strange psychology of non-Western peoples any more than the Western mind is reflected in the tales of Peter Schlemihl or of Alexander the Great's horse.

Some of the more bizarre "beliefs" of magic or religion have the common thread that they are strongly counterintuitive or that they somehow deny too strenuously the facts that are normally considered to be true. No culture gives any particular weight to the belief that grass is green or that the sun heats things up. But they do invest much ritual energy in immortality, or in the possibility of communicating with the dead or influencing other people's lives from a distance. In all these cases, when we say we believe in something, we are really showing that we think, or we know perfectly well, that things are quite otherwise: that there is no life after death, that the deceased do not speak to us, and that we cannot change our enemy's destiny by poking pins into an effigy.

Stories about shadows fall into this category. They seem, for example, to flagrantly violate our knowledge about causality and the material world, as can be seen from the case of the purification ritual required of the Brahmin contaminated by the shadow of an Untouchable, or the case of the cutoff shadow of Peter Schlemihl. But these stories are irresistible precisely because we all know perfectly well the workings of causality and materiality.

The Richness of Shadow

Shadows are seductive because they're so strange, and the language of metaphor has drawn abundantly from the wealth of images born of shadow. Shadows are immaterial, they have no substance, and for this reason a person who is *the shadow of his former self* only appears to be what he once was. One supports a *shadow government,* and it's useless to *chase after shadows.* In German they say you can't jump on your own shadow—in other words, you can't achieve the impossible. Hades was populated with the *shadows of the dead,* with bloodless duplicates of beings that were once healthy and whole. Shadows are parasites of the objects casting the shadows, whose shape they sometimes reproduce. They always seem to be relegated to the realm of *appearances.* A shadow is an image, a representation of the object casting the shadow. But it is an incomplete representation, a silhouette representing only an outline; the interior of the shadow is indistinct and it says nothing about the object casting the shadow except that it is an opaque, not a transparent, object. Someone losing his sight says that he *sees shadows.* A shadow is a *trace.* (*Skia,* the ancient Greek term for shadow, also means "trace.") One can speak *without the shadow of a doubt.* As an image, a shadow can also stand in for the object projecting it and become its double. Some shadows rise to the rank of real literary characters (and not only the shadows of Peter Schlemihl and Peter Pan: in Stravinsky's *Rake's Progress* the demon is called Nick Shadow). In English we are *shadowing someone* if we are tailing him: in this case the bodily shadow becomes *something that you cannot shake off.* (In medieval Arabic, by contrast, the name for a shadow is "the follower.") Because a shadow can hide an object, it is linked not only to the physical aspect of the absence of light, but also to the perceptual aspect, to visibility. Criminals *plot in the shadows;* other people are *relegated to the shadows* or are *brought out of the shadows;* a written text can have shadows, or passages that are not very clear; and sometimes to save a friendship you have to *clear away the shadows* that threaten it. Shadows can hide things because they're dim or even obscure. The German *Schatt,* like the English "shadow," derives from the Greek *skot-,* which indicates obscurity. In American Sign Language a shadow is a *black stain.* Obscurity can protect and it can dominate. You are *in someone's shadow* when you are protected by his power, but the protector can also *overshadow* whoever is close to him, keeping him from being seen.

The Moral of All These Stories

There is a lesson to be learned from all these linguistic, ethnographic, and literary reports, a lesson about the constellation of images gathered around the idea of shadows. The reports seem to say that we are *fascinated* by shadows; and it's not hard to see why. *Shadows do weird things.* A shadow is the image of a body, and it permits us to recognize its owner. It depends on the body—in fact, it's tightly bound to it. But it is also a ghostly, immaterial image; it is colorless and it is flat (it may be the only nonabstract object to be truly two-dimensional). Shadows grow and shrink; they appear and disappear; they are tied to their owners and yet they cannot be captured.

If we try to organize the rich metaphors about shadow, we can see that the concept crystallizes around two opposite poles. On the one hand are the phenomena belonging to "naïve physics" (the prescientific theory about the physical world); on the other hand are the phenomena of "naïve psychology" (the prescientific theory of the mental world). Naïve physics is made up of laws that roughly predict how objects will behave: an unsupported body tends to fall down, for example, and two bodies cannot occupy the same space simultaneously. Naïve psychology explains human behavior by resorting to simple theories that (for example) refer to a person's "character" as peaceable or choleric. Shadows seem to inhabit a part of the mind that opens onto the objects department—shadows are physical things—and also opens onto the psychic department—shadows are images of the soul. In reflecting upon the strange behavior of shadows, both departments are put to work. This duplicity is probably where we can find the cognitive explanation for the rich metaphors and stories about shadows.

Shadows bear witness to the meeting between the world of material things and a world in which matter does not seem so important. It is a capricious world—shadows come and go, without our being able to interfere with their plans—that is both evanescent and mysterious. It is a world of persons and of things whose chief features are basically the same across all cultures.

IV *Shadows on the Mind*

> If you hide my shadow with your shadow, mine still exists, because if you move away I can see my shadow.
>
> —Carlotta, nine years old

A very smart (or even a not so smart) child discovers his or her shadow at the beach at sunset (or playing near a long wall in full sunshine) and tries to jump on it (or to grab it). She follows the shadow, laughing with joy (or he follows it, crying with frustration) for several miles, ending up in the arms of friendly vacationers (or evil shepherds) who free her from the spell (or who kidnap him).

If you ever happen to write a book about shadows, you will find that there's no escaping the numerous variants of the story above. Sometimes the child's social station is different, or the time of year, or other elements, but it stands as a commonplace narrative born of the ambiguous spell woven by shadows. In the story the shadow is an object that cannot be controlled and that therefore has a sort of autonomous will. It changes from a simple optic phenomenon to an animate thing. The child is bewitched by the shadow. We must find the kernel of truth in this story; maybe the legend expresses nothing more than the charm, for adults, in talking about shadows being transformed into things and being endowed with a soul. In any case, the question is worth pursuing.

M, the Monster of Düsseldorf (a.k.a. Peter Kurten), and his victim, in Fritz Lang's movie M *(1931)*

Baby Shadows

Are shadows endowed with a life of their own? Children really seem to think so. Even if they won't admit it when pressed, they still may think that shadows are like real people—for example, that shadows exist even in the dark. But at what point do children discover shadows for the first time? And how can we know what children know about them?

Between the two world wars the psychologist Jean Piaget (1896–1980) started some interesting experiments on the world of the child, asking whether a child's reasoning is different from that of an adult and whether children treat reality the way adults do. Piaget's theories, published in a series of monographs over the course of half a century, are controversial and to this day are not generally accepted; but they have been unquestionably useful in promoting child cognition as a subject of research. According to Piaget, children's view of the world is fundamentally different from that of adults; they reify almost everything and they grant almost everything a soul until (like little scientists—by experimentation, hypothesis, and trial and error) they approach the adult view of the world and finally absorb it. We free ourselves intellectually from childhood at the moment when we replace a world of objects and animistic causes with a theory that emphasizes the fabric of relations between things. This process moves by stages in which the child's original concepts cohabit, to varying degrees, with gradually more sophisticated concepts.

Given that children tend to reify all kinds of phenomena, it would be interesting to see what children think of shadows. Shadows are part of the physical world (they're not like dreams) without being material: they are not typical physical objects. If the child associates shadows with some material characteristics, this means that reification is really a fundamental psychological mechanism.

But how can one understand what a child really thinks? Piaget believed that the best way was to ask the child himself. (Naturally, this works only if the child already can talk, so it excludes preverbal children.) And indeed, nothing could be simpler.

The children interviewed about shadows were between five and ten years old. Their answers during the interview were subtle and inventive. Gall (the strange names are codes used by Piaget) said that the shadow of a hand is dark *because the hand has bones.* Tab thought that shadow was created by night. Re was convinced that his shadow could be torn off by piercing it with an umbrella, and when he was shown that the shadow actually stayed in place, he replied that *there's another shadow* underneath the first one. According to Stei, the chair's shadow didn't move because it was caught under the chair's legs. In shadows Gill and Leo saw the portrait and the model of the person casting the shadow. Bab had an audacious theory: objects cast a shadow only on one side in the daytime because *it's nighttime on the other side.* So what happens at night? Well, at night objects cast shadows *on both sides.* For Roy shadow was a substance that occupied space and was impermeable to light. And there were also some who said that the shadowed side was dark because *the light can't see it.*

The comprehension of shadows proceeds by stages. In the first, most primitive stage studied by Piaget (starting at five years old) an object's shadow was thought to emanate from and depend on the shadow of the surrounding environment. Night and darkness are *black clouds* and daytime shadows are part of them. In the second stage (six to seven years old) shadow appears to be an emanation not of the night but of the object that casts the shadow. It follows that the child cannot predict where a shadow will fall in relation to a light source. For example, the child will try to run his shadow around the wall of the room by *spinning his own body* instead of walking around the lamp as he should. In the third stage (seven or eight years old) the child discovers that shadow has a geometric relationship with the source of light, but he doesn't yet see the causal relationship between the light projection and the configuration of the shadow; for example, he believes that an object makes a shadow even in the dark. In the fourth stage, at eight or nine years old, the explanation of shadows is purely geometric, and it is practically the same explanation that an adult has: that a shadow depends on an object that blocks a light source.

So at the beginning a shadow is considered to be made of some material—darkness—that turns out to be the same material of which night is made. In both the second and the third stages, shadow is a sort of object in itself; only in the fourth stage does this *shadow-thing* dissolve into the fabric of relations between objects and light. The crucial stage is the third one, a hybrid of the idea of the shadow-thing crossed with the correct understanding of shadows. The child cannot see the correct explanation because the idea of the shadow-thing is still too strong.

Battling with Shadows

All this is fascinating, but is it true? The psychologist Rheta DeVries reconstructed Piaget's experiments with a broader sample of children, and she rewrote the story of the development of shadow. Her results partly confirm and partly refute Piaget's hypotheses. For example, her sample children really consider a shadow to be an object, an emanation. But it's not part of the night; when a shadow disappears in the dark, it's not joining with the great nighttime shadow—it's *hiding within each body.*

The experiment is done during a game in a room with a movable lamp, which makes shadows on the wall. The experiment leader and the child have things that cast shadows, and the leader tries to get the child to do certain tasks: to make her shadow larger than that of a toy; to make her own shadow touch the leader's shadow; to move the shadow or make it go away. The child must answer many questions: When my shadow touches yours, does it have the same effect as when I tap your shoulder? Where do shadows go at night? Can you keep your shadow with you? (The questions are cleverly designed to establish whether the child has a coherent idea of what she's saying.) DeVries, like Piaget, found that the idea of shadow changes over time. At level zero there is

Does shadow beat gravel? Three post-Piaget children trying in vain to make a shadow disappear by covering it with pebbles.

almost no idea of the relationship between an object and a shadow. At the first level the shadow coordinates with an object because it resembles it, or because it shows up near the object; but the child still doesn't grasp how to control it. Little Julia, for example, would like to see the shadow of her wooden horse's tail, but instead of setting the toy sideways across the beam of light, she points the tail at the wall, as if the shadow could spurt out from the object.

At the second level, lighting comes into play, tangled in paradox: how can light produce something dark? One boy solves the problem in his own way by saying that *the shadow is shining on the table.* In this stage, light is considered to cause shadow, but the geometry of it is still confused. For example, even in a sealed room the sun is mentioned as a cause for the shadow. But the situation changes quickly when the children begin (by trial and error) to understand the importance of the relative positions of the lamp, the object casting a shadow, and the screen. For little Aaron, this apprenticeship phase was particularly hard. When asked to cast his shadow on a wall other than where it was (a task requiring that the lamp be moved), he tried to drag the shadow by pulling it up off the carpet or pushing it with his feet. Aaron was battling with his shadow. Soon children realize that they can control distant shadows, and they understand better the role played by light. The results can be quite creative. Kate, at seven, imagines that the light makes a shadow because it wraps around her hand; Ann, nine, considers light to be a kind of wind that carries shadows with it. At the third level this ability becomes geometrically more sophisticated, but it's not based on a comprehension of the phenomenon of shadow. The *how* of shadow is understood, but not the *why*. Children can predict where the shadows will be projected, but they think that their shadows continue to exist even if they're covered up by the experiment leader's shadow, or even in the dark (light serves only to *reveal* a shadow *that was already present* in the dark). The transitory nature of shadows is grasped only at the fourth level, the level that completes the assimilation of childish shadow concepts to adult concepts.

Immaterial Material

But not everyone agrees on how children think about the materiality of shadows. Some researchers (Carol Smith, Susan Carey, and Marianne Wiser) grew suspicious when they asked children whether shadows *are made of something*. The answers were almost always affirmative, but the children often added that shadows are made *by you and by the sun*, meaning that they probably interpreted the question as if it referred not to the physical composition of shadows, but to the process that generates them. The researchers therefore used a method different from direct questioning. They explained the distinction between material and immaterial to the children (between four and twelve years old) by saying that some things, like rocks and animals, are made of matter, while others, like ideas or sadness, are not made of anything. At this point they presented the children with a list—car, tree, water, Coca-Cola, idea, electricity, light, shadow, echo, dream—and they asked the children to sort through it and to divide the material things from the immaterial. Starting as early as four years old, all the children put shadows in the immaterial zone, together with echo, dream, and electricity. This refutes the thesis of Piaget and DeVries: children do not think that shadows are material.

If we accept that children consider shadows to be objective, to be a part of the external world, we can conclude that the concept of an objective thing is not linked to materiality even for the youngest children. Children are not as materialist as Piaget would have them be. In any case, their concept of matter is complex and does not quite match the (prescientific) concept in adults. According to Carey, until the age of twelve, children consider immaterial certain things (like air) that we consider material, and they consider material certain things (like heat) that we consider immaterial.

Up to this point we know only what children tell us or what we can extract from the contradictions in their answers. Their thinking about shadow might derive from adult conversations they have heard, or from experiments they made while playing; or could it be a case of unlearned concepts, linked to an earlier phase of their childhood development? To discover the answer it would be interesting to question preverbal children. But with infants who can't yet talk, how can we know what they think about shadows? And not just about shadows: how can we know what children think in general?

The answer is: you have to *make them bored.*

Boredom as a Tool for Knowledge

On your right is a great dark spot that grows and shrinks according to an unpredictable rhythm. It is accompanied by a modulated sound, like a record on a phonograph that's winding down, turning slower and slower. You're frozen with terror, partly because you're also swimming in a colorless, indistinct liquid where there's no up or down, and the threatening dark spot can grab you at any time and hold you tight. Every now and then a cloud of perfume wafts by; for that brief moment you are the cloud—there's no difference between you and the cloud, just as there is no difference between you and the sound you're hearing, and maybe there's no difference between you and the dark spot. But now the spot seems less disturbing: it's moving toward you with music, with a song. A curtain is pulled back, and all of a sudden you see the soft object of your desire.

This is not the start of a science-fiction novel, but rather the description of a baby's-eye view of reality right before he nurses. It's an intriguing picture; too bad it's completely false. Descriptions like this are the scientific equivalent of urban legends that have long since been repudiated by experimental research. The legend dies hard, though, because people think, Well, we'll never know what really goes through the mind of those little rugrats. But this too is false.

Since the 1970s, psychologists of infant development have used a technique known as the habituation method, which permits them to study the mental universe of preverbal children. (The studies start as early as the first few weeks of life, but we'll limit this discussion to babies already a few months old.) The basic idea is simple. To start with, you have to bore the baby by repeatedly presenting him with the same situation—for example, three sounds, and then the same three sounds again, and then again—for a long time. At the beginning the baby is interested (he sucks more vigorously); then he's less interested; and ultimately he gets distracted: he's bored by it. When the vigor of sucking drops off and it's clear that the baby is completely bored, the researcher moves on to the test situation by presenting *two* sounds. The baby snaps back to attention. How can we explain this interest on the baby's part? The only change in the situation is the number of sounds, so we can hypothesize that newborns can tell the difference between two and three. Or rather that they have some idea of number and quantity. The conclusion is interesting: at four months, babies are sensitive to differences in quantity—even if it's only up to three signal noises. (The babies react with less enthusiasm to a switch from three to four stimuli.) The experiments are long and complicated, and the researchers must surmount enormous obstacles: it's not easy to find the infants; they need to be carefully selected. And the parents must be willing to hold the babies in their arms if they are fussy; the parents must also shut their eyes while the experiment setup is presented (to avoid influencing the babies' responses by transmitting signals with their bodies). But it's worth all this trouble.

Working with legions of bored babies, researchers have discovered that the babies know a lot of things that they cannot have been taught: they can, for example, add and subtract small sums; they know that objects continue to exist even after they disappear behind a screen; that different objects cannot occupy the same space at the same moment, and that they move along continuous trajectories without unexplained leaps. Furthermore, babies don't accept the idea of action at a distance: they believe that two surfaces move in coordination only if they are in contact with one another; they believe that movement is transferred in the collision between objects; they are surprised when there is a discrepancy between tactile stimuli and visual stimuli, so they are able to compare information coming from different senses. And so on.

What babies know is interesting because this news disproves a certain long-rooted idea about the nature of learning in infants. It's not true that a baby lives his life shut in a narrow subjective prison, in a mushy, jellylike world of unfocused color and unsettling sounds, with erratic flashes from hunger pangs or powerful odors, a world brightened only by occasional tasty moments of nursing, the single great event that's the only thing worth learning to recognize.

Quite early on, a baby sees the world more or less the way we do: colors are not subjective but real; things behave predictably and they interact with other things in a network of cause and effect that obeys precise laws. The world of very early infancy is significantly richer than it seems: it's not a simple mosaic of sensations, for it includes the causal notion of contact and the metaphysical notion of the permanence of unseen objects. Indeed, one might conclude that newborns have a theory of the world, or maybe a series of minitheories, one for each type of object. Of course, their theory of the world differs from our theory; but these differences are exactly that—differences between *theories* and not between an indistinct perception (which would be theirs) and a structured perception (ours).

Among these minitheories, is there one about shadow?

Shadow, Are You the Thing?

As for Piaget's and DeVries's older children, shadows are for preverbal children a wonderful litmus test for the conception of physical reality. Shadows are physical objects, but they are immaterial, so they need a place of their own in the collection of theories about the world. But for very young children there is another problem. Just look around: shadows are very prominent because they are dark, high-contrast spots in the visual field. A pebble on a sunny beach casts a shadow on the sand far darker than the stone itself. So a shadow risks *confusing* a newborn, who might take it for a material object and invest too many mental resources in it. To solve this problem, evolution might have designed babies who just *did not see* shadows, providing, for example, a cognitive "filter" canceling them out. An experiment by Gretchen Van de Walle, Jayne Rubenstein, and Elizabeth Spelke shows, however, that babies (between five and eight months old) recognize the difference between a shadow standing still and a shadow moving, and that they thus register the presence of shadows rather than filtering them out completely, though it is not clear whether they see them *as shadows* or as dark spots.

In another experiment our researchers bore babies by showing them a sphere that stays suspended for a while, casting a shadow below it onto a box.

At this point they try to understand whether the babies find it more

The box, the shadow, and the sphere can each move independently. Experiment 1: In an unnatural situation, the shadow moves while the sphere stays still; in a natural situation, the shadow moves with the sphere. Experiment 2: In an unnatural situation, the shadow moves with the box; in a natural situation, when the box moves, the shadow stays still beneath the sphere.

surprising when the shadow moves (unnaturally) while the sphere stays put, or when the shadow moves (naturally) to follow the sphere as it moves. Contrary to what happens with adults, the babies are more surprised by the natural movement of the shadow—that is, it pleases them less. Why? One hypothesis is that the natural movement of the shadow violates a fundamental principle of baby physics, which prescribes that *there are no actions at a distance.* The shadow should not move because it is *attached to the box* and not to the sphere.

A further experiment seems to confirm this data. The baby is made bored as in the previous experiment (he is shown the sphere casting a shadow on the box, with sphere, box, and shadow all immobile) and then one tries to discover whether he is more surprised by the box moving while the shadow remains immobile beneath the sphere (a natural situation) or by the box moving and the shadow, glued to the box, moving along with it (an unnatural situation). Again, it is the natural situation that most surprises the baby. One could conclude that the box moving with the shadow staying in place—as it should—clashes with another infantile theory of the world: *if there is contact, there is (inter)action.* Contact and action at a distance are causal links: if I see an ashtray on a moving box, I expect the ashtray to move with the box. If I see an ashtray here and a box over there, I don't expect that the ashtray will shift when the box is moved.

Babies treat shadows somewhat like the ashtray, and it would be interesting to understand why they do, because shadows are *not* things like ashtrays.

One very simple hypothesis is that babies "respond" to shadows because shadows are conspicuous, so automatically—without thinking twice—they apply to shadows the principles that are valid for other objects. Shadows are treated as material things. Why? It doesn't cost much to do so: babies already have a theory about material things.

But there is one last curiosity. When babies see the shadows—which they consider objects—*doing things that would be strange for an object* (sliding across a box, or coordinating their movement with distant spheres), what do they think? Do they see them as shadows? Do they learn to treat them as shadows? Or are they completely confused by this?

Van de Walle, Rubenstein, and Spelke have built a subtle conceptual trap for the babies. First they got them used to situations that the babies initially found surprising, showing them a shadow that does not move with the movement of the box it's sitting on, or a shadow that moves together with the object casting it. Then they tried to see whether the babies were truly accustomed to this new kind of object; and the answer is that the babies really don't know what to think anymore.

Even if the interpretation of these experiments is not yet completely certain, it does seem that babies in their first months of life do *not* have a true *theory of shadows* that comes into play when they understand that shadows don't behave like objects. They go on treating them like objects when they can, and they err in their expectations of how shadows will behave. Of course, with time and experience they understand shadows: children learn to live with shadows, and they refine their conception until it matches our adult conception.

The struggle has a rational outcome. Unless, that is, the adults too are subjugated by shadows.

V *Shadow of a Doubt*

"Mary, Mary, why didn't you keep that shadow?"

—Mr. Darling in *Peter Pan*

The world we live in seems fairly simple, but astrophysicists and quantum physicists have gotten us used to bizarre entities and events whose behavior is hardly intuitive: black holes, spatiotemporal singularities, quarks. Science seems to force us to abandon the solid certainties of our everyday experiences—a world made up of rocks and tables and chairs—with reports that we live in a zoo full of strange creatures. But there's no need to look at big bangs and subatomic particles to feel lost. The world of normal experience is worse than a zoo: it's a jungle hiding metaphysically suspect creatures. Just consider a *hole* in Swiss cheese, the *beauty* of a flower, the *number* two, Beethoven's *Fifth Symphony,* the Mona Lisa's *smile,* Mike Tyson's *fists.* What is a hole or a smile? Could we define them? What is a fist? Is it an object different from a hand? Let's add shadows to the list of suspect things. Shadows, which are always in view, do odd things that strike the popular fancy, but they also do even odder things that give us food for thought.

For one thing, shadows are reciprocally penetrable: two shadows (made by two separate lights) can occupy the same space without bothering one another. Or a shadow can split up and still be the same shadow even if its pieces occupy spaces that are not attached: the shadow of a statue can fall partly on the ground and partly on the table the statue stands on.

Two shadows stuck together don't make one shadow, and splitting a single shadow doesn't result in two shadows.

Two distinct shadows can unite and give the impression of a single shadow while still remaining distinct: the part of the statue's shadow that lies on the ground melds into the table's shadow on the ground, but even without a dividing line they are two separate shadows. And if we turn off the light and then turn it on again, we can't be sure of getting the *same* shadow as before.

What is the touchstone that lets us say that something is more or less counterintuitive?

It's a stone, that's what.

Rocks (and all other *material* things) serve as our model for things that behave "well"; certainly they behave *very* differently from shadows: two rocks cannot occupy the same space, and a rock is no longer the same rock if we smash it to pieces. If we meld two rocks together, we create a new, third rock, and the other two disappear. A rock seems to move through space in a continuous manner. If we leave a rock somewhere, we come back to find it in the same place; and if we cannot find it, we know that's because someone moved it.

Shadows are immaterial things and are therefore quite different from the material objects of everyday experience (which are the very things that cast shadows). How different? We can see how different if we make a list of *shadow brain-teasers.* Brain-teasers are serious *divertissements.* They show how problematic are the concepts that we use without any difficulty to describe the world around us.

Brain-Teaser Number One: Shadow or Night?

For example: We take shelter under a tree at midday and we say we're in shadow. Night falls, we walk out beneath the open sky, but we don't feel as if we're in shadow. And yet isn't it true that we're in the earth's shadow cone? The symbol for night in one of the earliest known written languages is shaped like a celestial vault with darkness descending from it.

Night, in Mesopotamian Oruk writing, third millennium B.C.

Night is thought of as an exclusive property of the sky and not as the consequence of the earth's interference. While night seems not to be part of the concept of shadow, that's only because the concept is not *elastic* enough and, indeed, on one fundamental point it's absolutely rigid: it doesn't recognize the presence of shadow if it doesn't find at least some light. That is to say, *there is no shadow if there is no shadow line*—if there is no separation between shadow and light. After sunset this separation disappears, and with it disappears the possibility of treating night as a shadow.

We don't see the great shadow of the earth projected in the sky during the night because we don't see the light that fills the night sky. We are able to see the presence of light only when we see an object that emits it or reflects it. Often there is a lot of light but it cannot be seen. Observation of the night sky proves this: it is a bottomless well and it doesn't seem at all to be crossed by a great swath of light emanating from the sun (as is indeed the case). If the rays of the sun were visible like glowing filaments—which would be physically impossible: the filaments would have to emit their own rays of light!—we would see the night sky crisscrossed with them all over. The light that fills the night sky is revealed to us only by interplanetary dust in suspension (this very weak glow is known as zodiac light) and by the moon and the planets that intercept the light and reflect it toward us.

Brain-Teaser Number Two: In the Shade or in the Shadow?

Some languages have two distinct words to distinguish the concept of shadow itself from a shadowed space: English, for example, uses "shadow" and "shade," while Italian doesn't use two terms. (It does use the single term *ombra* in two distinct ways, recognizing the difference between the two concepts.) You can stand in the *shade* of a tree or in the *shadow* of a tree. In the first case the shadow is an indistinct volume of darkness; in the second case it is a well-defined object. This is demonstrated by a simple grammatical reflection. If the tree makes two shadows, I can stand *in one* of the two shadows, but grammatically it makes no sense to say that I can stand *in one* of the two shades. When you can use a numeral before a word (there is *one* shadow), you're usually talking

The figure is in the shadow of the tree, without being in the shade of the tree.

about objects (there is one table, one chair). But when you can use a partitive (there is *some* shade), you're usually talking about stuff without spatial organization (there is some milk, there is some sand). Thus there are two diverse (although related) notions.

Brain-Teaser Number Three: Body or Surface?

Shadows seem to be a good example—perhaps the best example—of a two-dimensional object. No matter how thin, a sheet of paper still has some thickness. When Edwin Abbott (1838–1926) wished to describe the characters in the flat world of *Flatland* (1882), he compared them to shadows. But we don't always treat shadows as two-dimensional objects. Grab a brick and hold it over a table beneath a lamp. The brick casts a shadow. Now lay the brick on the table. Does the brick still cast a shadow on the table? Maybe not: making a shadow seems to require some space and thus excludes contact between the brick and the table. And what can we say about the *inside* of the brick? Is it in shadow or not? After all, shadow is the absence of light, and inside the brick there is no light. But the concept of shadow doesn't seem to go all the way to the heart of objects.

Brain-Teaser Number Four: Does a Shadow Survive Being Swallowed?

There is a magical moment when the shadow cast by a mountain seems to swallow up my shadow and those of all the trees in the valley. What happens from that moment on? Piaget's children held that shadows exist even in the dark; adult cognition seems to hold, instead, that in the darkness both my shadow and those of the trees literally exist no longer. Who is right? Children take the easy route: if the shadow doesn't disappear, it will be the same tomorrow when the sun comes up. Maybe adults cling instead to the principle that there's no shadow when there is no shadow line separating it from an area that is lit.

Brain-Teaser Number Five: Is it the Same One or a Different One?

Perhaps Piaget's children are right. I turn off the light when I leave the room and I turn it on again when I come back in a moment later. Is the shadow of the statue on the table the same one before and after the light is off? If we say it's the same, then we accept the existence of strange, intermittent entities (in the same family as the flashing of the turn-signal light on a car). If we don't like that idea, then we have to say that there are two different things: the shadow that was present before I turned off the light and the shadow present after I turn it on again. What's different about them? The lamp, the table, and the statue are the same before and after (or we could decide that they are, so as not to complicate things). So there must be some other factor that determines the difference. For example, the emission of light. This is a process that certainly has different phases in time: the photons emitted by the lamp before I left the room are different from those emitted after I come back. But there's a problem here: new photons arrive every instant, so if the identity of the shadow depends on the identity of the photons, the statue creates a new shadow on the table much more often than we would like—every instant there's a new shadow! From this we must conclude that although shadows seem to be entities with a life in time (let's say, the shadow cast by the statue between six and seven o'clock), this is just an illusion, a mental simplification. The shadow between six and seven is really the sum of innumerable instantaneous shadows.

So we face a dilemma: either we accept that shadows are intermittent, or we must accept the idea that their persistence is just an illusion.

Brain-Teaser Number Six: Inert or Active?

With a hammer I can drive a nail into the wall, but with the shadow of a hammer I cannot drive a nail. Shadows cannot interact with material things and *not even with other shadows.* (With the shadow of a hammer I cannot drive the shadow of a nail, not even when we're talking about the shadow of a hammer at the moment that the actual hammer is driving a nail!) Shadows are *inert.* At times it can seem that this isn't true. On the beach I make a shadow with my hand. If I'm asked whether the sand is cooled by my hand or by its shadow, I would say it's the shadow. This preference seems to be linked to a principle of intuitive physics (shared, as we have seen, by children): intuitive physics denies that there can be action at a distance—the hand cannot cool the sand because it's not touching the sand. But shadow has this power because it is in contact with the sand. All this, you might reasonably say, depends on the fact that common sense hasn't got its ideas straight about how cooling works.

Brain-Teaser Number Seven: Is Shadow Faster than Light?

Shadows seem to be static, but we know that the light which defines them does move. There is a cat toy that casts the shadow of a mouse; cats can spend hours hunting this intangible mouse. The shadow is made by affixing a little mouse-shaped sticker to the lens of a flashlight. If we point the flashlight ray toward a star, we create a hole in the light vaulting up into the sky. If we remove the mouse sticker, this shadow ceases to exist down here, but the light now follows the back edge of the mouse-shaped shadow across space. The ray of light continues to voyage through space, taking with it the shadow that moves along toward the star at the speed of light.

Paradoxically, a shadow can go even faster than light! We can turn the mouse flashlight to point first at one star and then toward another star. We can take two stars more or less the same distance from Earth, such as Acrux in the Southern Cross and Bellatrix in Orion (both are 360 light years away from us). We point the flashlight at Acrux and then we make the beam slide slowly toward Bellatrix. Three hundred sixty years later, the shadow of the mouse (which will at this point have become huge and very fast) will reach Acrux, and just a few seconds later it will be at Bellatrix, after having crossed one quarter of the vault of the sky far faster than the speed of light. Can shadows do things that are physically impossible?

Eight: The Great Brain-Teaser About Shadow, or, Too Many Causes Spoil the Effect

All these problems join together in the Great Brain-Teaser of shadow, described by the philosophers Samuel Todes and Charles Daniels.

If we wish to understand what shadow is by following our common sense, we must act like Sherlock Holmes and begin gathering some elementary facts about shadow to see whether we can extract a little theory. We quickly discover that if there's a shadow around, there must also be a body somewhere making the shadow. That is, shadows *always depend on an obstacle blocking the light.* (As directors of horror movies know; they have the villain come on the scene preceded by his own shadow. Seeing the shadow, we know that the shadow's owner, alas, is somewhere near.)

There's more. We also know that an object *does not cast a shadow through another object.* The table makes a shadow on the terrace. If I put a statue on the table, the statue makes a shadow on the surface of the table. But this shadow does not "pass through" the table to reach the terrace floor. (Try it and see.)

What else do we know? Naturally, we know that to make a shadow a body *must be illuminated,* let's say from one side. If it doesn't get light, the table won't make a shadow.

A really minimal theory of shadow must therefore contain at least these three principles: one, every shadow is the shadow of some body; two, a body doesn't cast its shadow through another body; three, to make a shadow a body must get light.

Now, it's possible for even such a simple theory not to function. There is a very basic situation in which this theory would be utterly confounded.

It is the following situation. Let's go back again to the table casting its shadow on the terrace floor. Let's take the little statue and put it *under* the table, in the shadow, in such a way that its own shadow does not poke out from the shadow of the table. We move the table away for a moment and—still like Sherlock Holmes—we trace a chalk outline of the statue's shadow on the terrace. Then we put the table back in place, leaving the statue where it is. We now have the statue back under the table, and inside the table's big shadow is an area marked off in chalk. This chalk line encloses a shadow zone that we'll call the Suspicious Zone. Now we ask: *whose shadow* is in the Suspicious Zone?

A little statue stands in the shade of the table. The white dotted line marks the Suspicious Zone. What is it the shadow of? Of the statue or of the table? Neither answer sounds good.

It's not the table's shadow, since according to the second principle the table cannot cast its shadow through the statue. It's not the statue's shadow, because according to the third principle the statue is not illuminated, standing as it is in the table's shadow. But beware: because there's nothing else that can cast a shadow besides the table and the statue, it seems that the first principle is useless here. The Suspicious Zone is a shadow, but it's not the shadow of any one object.

Something is not right with the theory of shadow, even though it seemed so simple!

The problem is not as philosophical as it seems: it could have interesting legal implications. Let's recall Tokyo's antishadow rules.

A little house sits in the sunshine until the day that a developer begins constructing two buildings, Big and Small. The developer designs them so that the roofs of Big, Small, and the little house are all lined up in relation to the sun. In this way he gets around the rule: Big isn't shadowing the little house, because its shadow doesn't go beyond Small; Small isn't shadowing the little house, because Small doesn't get any sunlight. And so the owner of the little house gets no compensation.

To get around the antishadow laws, it's best to build *two* skyscrapers instead of one.

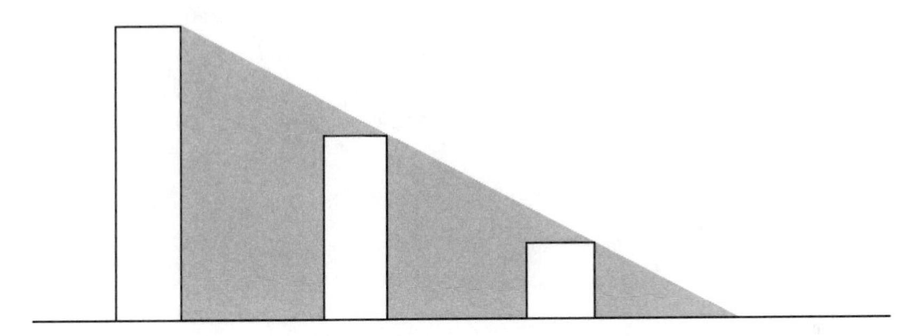

How to use metaphysics to avoid paying taxes

The Solution and the Hole in the Shadow

It's a good rule of thumb to never propose a problem without suggesting a solution. So here it is. The brain-teasers arise from the fact that shadow is considered to be a material thing with a shape of its own. Only when we forget that it's not material can we say that shadow travels faster than light (Brain-Teaser Number Seven) or that it cannot pass through another body (Brain-Teaser Number One and the Great Brain-Teaser). Our minds obsess over concrete things and cannot understand what happens when one shadow is swallowed up by another, when shadows disappear and reappear, or when they seem anything but inert (Numbers Four, Five, and Six). Thanks to light, shadow survives, and it is therefore not only immaterial but also dynamic; when we forget this, we cannot grasp how shadow can be both unitary and structureless, flat and three-dimensional (Numbers Two and Three).

The brain-teasers show that the concept of shadow is a *hybrid*, put together by a sloppy do-it-yourselfer: it has a *causal* aspect, a *material* aspect, and a *perceptive* aspect (Brain-Teasers One through Five, and the Great Brain-Teaser: we need to be able to *see* shadow, to see the shadow line, in order to say that there is a shadow). It's impossible to correct this tendency of the mind to treat shadows as things with a figure. Look at this photograph:

The image of the sun during an eclipse appears amid the shadow of interlaced fingers. These are holes in the shadow.

The shadow is full of holes. If there were no shadow, there could be no holes in the shadow. This is completely obvious from a perceptual point of view: the shadow is like a material support for the hole. If we treat shadow like a thing, it's because shadow always draws us into doing that.

The Shadow-Object and the Closing Surprise

The moral of the story is this: common sense wavers, at times considering shadows to be objects and at other times considering them as fuzzy noncorporeal phenomena.

Now the shadows of science are in a similar position.

The great *shadow cones* that astronomers mention when discussing eclipses are objectlike shadows: they have shape and dimension, and they are geometric bodies in every way, just as cones of rock would be. Above all, they are geometric bodies that are relatively easy to imagine and to study. A punctilious astronomer might object that shadows are not material objects (that they are in fact absences). Driven by the need to be orderly about the concepts he uses, he might rewrite the dynamics of an eclipse by *discussing only light* (which has a less uncertain material state, and which, in any case, is not an absence). But there is a problem:

at this point shadows would play a role as *holes in the light*, which naturally pushes the problem away only slightly, insofar as holes are just as immaterial as shadows—they too are absences. (Holes are like shadows in many respects: for example, they don't have internal structure. They are entirely internal, without a fabric of their own. If an object with a round hole spins on the axis of the hole, does the hole rotate too? If an object that makes a perfectly circular shadow spins on its axis, does the shadow rotate too?)

The most punctilious of astronomers, one who really disliked shadows, could describe an eclipse in apparently much stricter terms, without holes or other strange absences creeping into it; he could limit himself to describing the geometry of the flow of photons. But the shape of space occupied by light is *very* complicated to describe. Shadows stand in as a convenient conceptual shortcut. An analogy with the case of the hole shows why. If you want someone to cut a star-shaped hole in a piece of paper, the easiest way is to ask her to cut a star-shaped hole. The star is the shape of the hole, not the shape of the paper or of part of the paper. If you intend to punctiliously avoid mentioning holes because you don't believe they exist, you could give a long description of the internal shape of the paper, saying, for example: "You must shape angled inlets, with cuspid peaks between each pair." I doubt that you would succeed in making your friend cut a star-shaped hole, but even if you did succeed, the mental cost of giving that description is just too high.

Something similar happens with shadows. Scientific communication requires precise descriptions, and all descriptions cost something to the brain of whoever is doing the describing. It seems only natural that ever since the beginning of astronomy shadows have been treated as objects with their own geometry and their own autonomous existence, and that people never even began talking about the deeply complicated shape of the light surrounding them. It costs much less to do it this way.

The slightly odd conclusion to be drawn from this discussion harmonizes perfectly with the surprises that shadows force on us. A scientific discipline such as astronomy that accepts shadows is not materialist—because shadows are immaterial. So it's an error to associate science with materialism.

Part Two SHADOWS IN THE SKY

(Curtain rises)

Plato and His Shadow

Walking down to the Piraeus, Plato stops at a bend in the road and stands with his back to the sun. At his feet Skia, his shadow, is drawn particularly sharply on the ground. The wind ruffles the foliage of the olive trees.

SKIA *(desperately):* Shut up in a cave; used as an example of inferior knowledge; accused for centuries of being a scarecrow for philosophy. I'm going to say it again: I keep you company all day, from sunrise to sunset, and you just walk all over me. You should beg my forgiveness.

PLATO: How impertinent! You've seen for yourself that you're not only ephemeral and dark, but a nest of contradictions; you create confusion and fear, and you puzzle children and adults alike. It seems to me that your position has only grown worse.

SKIA: But that's exactly the point! I want to prove to you that even though I'm scary, and even though humans don't know what to think of me, I can still be helpful to everybody—including scientists and philosophers like you.

PLATO: Oh, is that so? I really can't see what use you might be.

SKIA: Well, to start with, if it weren't for me, there would be no alternation of night and day, you wouldn't be able to see the shape of things, everything would look flat and insubstantial. . . .

PLATO *(exasperated):* Well, OK—but you're just a bit player. All the real work is done by the light.

SKIA: I object to that! Light only understands how to move straight along its same old path. That brainless fellow just shoots around without thinking, and when it comes up against some object, it bounces off and starts moving in another direction. *I'm* the one who preserves the traces of that meeting. Shadow is the memory of light. But you're still not convinced? Well, I have a lot of arrows left in my quiver. Read on and you will see.

VI Special Effects

> Light not its own, shining at night, wandering around the Earth with its gaze turned always toward the rays of the sun.
>
> —Parmenides, *Fragments*, 14, 15

Plato is unfair to shadow. A science such as astronomy couldn't survive without shadow. There are even some who think that rational philosophy was born when shadows revealed a valuable secret about the nature of the cosmos. One of the fundamental texts of Western thought, Parmenides' *Poem* (fifth century B.C.), may be no more than a philosophical flourish on an apparently banal observation: that the phases of the moon are just a shadow spectacle. The moon doesn't change—its mutations are just an illusion—and when one says that the moon is *waxing* or *waning*, one is merely succumbing to the theatrical tricks of shadow. Doubt begins to creep in—maybe all of reality is like the moon? Might it be that the changes in our own lives—life and death, the flowing of rivers, the changing of the skies—are just *special effects*, like the phases of the moon?

This is quite a thesis, and the simple example of the moon is not enough to justify it. So Parmenides tried to *demonstrate* that change, *any* change, is illusory. He did it by building an argument more or less like this: Given that only being exists (only being *is*), there is no nothingness and no void. So the world is full; but in a full world there can be no movement and therefore no change either. Change doesn't exist; it's just an illusion. This reasoning is a bit of a stretch; for the moment let's say merely that Parmenides' discovery of the shadow of the moon may have so beguiled him that he invented philosophical reasoning.

The fact remains that the shadowed part of the moon has revealed something we didn't know. The phases of the moon are illusory changes because they are nothing more than a display of shadow, seen from different points of view as the moon rotates around the earth. But where does this shadow come from? Beneath Parmenides' veil of poetry is hidden a surprising bit of astronomical information: the moon gazes at the sun and drinks in its light. (Karl Popper, one of the great philosophers of science of the twentieth century, declared that he was spellbound all his life by this revelation of Parmenides'.) How can it be that such a small idea is so astonishing? The moon is right in front of our eyes. It's a topic of conversation in every culture. Sooner or later, someone must have noticed that it is lighted by the sun. And yet this awareness wasn't even part of common knowledge—it's not included in whatever commonsense astronomy may have been handed down from generation to generation.

My point is that there's nothing at all banal about this knowledge. It is a real discovery that each of us has to make over and over again—with a certain feeling of astonishment each time.

So Large That It's Invisible

The reason for this astonishment is that the moon and the other heavenly bodies are objects too large for us to think about. It's easy to ponder a sphere the size of a billiard ball; it's much less easy when the sphere is more than 1,800 miles in diameter, and when it spins overhead at almost 2,500 miles an hour.

When we want to show the limits of our perception, we usually mention microscopic items. We cannot get very close to the subjects, and our eyes have insufficient power of resolution to distinguish between two dots no more than a pinpoint apart. But there is an opposite limit, for things that are too big—a limit based partly on the nature of perception and partly on broader cognitive facts. We cannot understand the shape of the earth because we cannot see it in its entirety; but even perfectly visible objects like the sun or the comets crossing the solar system are not easy to understand. The things themselves are not hard to see. The limit is cognitive: these things are too big and too far away to be grasped by the mind.

And so the mind presents these objects as best it can. Here's a simple experiment regarding our memory of the apparent size of heavenly bodies; this experiment should be done without checking the sky (the solution is below). Hold this book at arm's length and look at the drawing of the thumb. Compared to that thumb, how big should the moon appear to be? (And how about the sun?)

From memory it's hard to estimate the apparent size of the moon. If you stretch out your arm, how big does the moon seem, compared to your thumb?

Many readers will be surprised to hear that the apparent size of the moon corresponds more or less to the smallest of the moons in the series. For readers who are not convinced this is true—even readers with very long arms—here is an even more stunning demonstration. Take one of those hole-punchers for loose-leaf pages, make a hole (which should be ⅕ inch across) in the margin of the page here next to the series of moons, hold the page at arm's length, and look at the full moon through the hole.

In memory, the moon expands. But apparitions in the sky can be particularly tricky even when they don't pass through the filter of memory, which is notorious for its treachery. Let's consider a remarkable special effect, the so-called moon illusion. The moon seems much larger when it's near the horizon than when it's high in the sky, but this is a trick: if you measure it in both positions using the hole in the paper, you won't find any difference. (This holds true for any single day; there are small variations from day to day, due to the moon's differing distance from the earth at different points in its orbit.) Various explanations for this illusion have been proposed, of which the following seems to be the simplest. Space between here and the moon at its height is without structure and there's nothing for our vision to latch on to. But the space between us and the horizon is measured out with numerous obstacles—houses, trees, mountains—that offer a sense of depth. At ground level the brain creates a space in which to set objects, and the result is that the horizon always seems farther away than the zenith. The moon covers the same quantity of sky in the two positions, but the brain recalculates its size as a function of distance: since the moon seems farther away at the horizon, it also appears bigger.

Another factor influencing our ability to build a suitable idea of celestial bodies is linked to the subtended angles of those bodies in respect to their starry background or the foreground of earthly objects. These are known as *parallax* phenomena. When I see a tree hiding a distant house in front of me, I can move to my right and I'll see the house as being to the right of the tree; if I move left, I can see the house at the left of the tree. The parallax makes distant objects seem to follow me as I move. We all know what happens when we walk along a row of trees. The landscape elements like distant houses and mountains seem to follow us, but sooner or later we will leave them behind. The moon, on the other hand, never stops following us. The brain knows this simple rule: the more a landscape element follows us, the farther away it is. So the moon must be very far away.

But how far? It's impossible to say. In practice, when we look at the moon or the sun, we don't get any notion of their size or their distance. Maybe they appear bigger than the mountains that they're setting behind, but this information doesn't get used when we see them high in the sky. There is no perceptual cue as to their real size.

Natural Astronomy and the Ecology of Perception

Why have we taken this digression into perception? And what does it all have to do with shadows? The answer is simple: The earliest astronomical discoveries were made with the naked eye, and perception has its limits when dealing with celestial bodies. Shadows can carry us beyond those limits.

In other words, we have to take into account our cognitive mechanisms when we try to reconstruct the first great astronomical discoveries. We know that a naked-eye astronomy, which we might call *natural astronomy*, precedes instrumental astronomy. Its results are all the more surprising when you consider the limitations of our cognitive systems. The methods that were used to probe the cosmos with the naked eye are ingenious, and to appreciate them we must weigh what information could once be obtained (and what can be obtained even now) only on the basis of what could once (and can now) be seen. We know that the people who turned their gaze to the sky two thousand years ago had more or less the same ability to discern the stars, and that they reasoned more or less as we do today.

To understand the extraordinary intellectual achievement represented by their comprehension of the sky's structure, we must ourselves try to practice natural astronomy, forgetting all we know about celestial bodies and trying to interpret the bodies' movements as we see them. I have a fairly clear memory of the moment I began observing the sky with my naked eyes, without looking at any diagrams, trying only to remember how I might see in the night certain phenomena that I had read about. I was enormously frustrated, but I believe that's quite normal. Evolution didn't shape us to understand the stars: the books of natural astronomy ought to be written by psychologists.

To see how far astronomical knowledge succeeds in permeating human cognition and common knowledge, I joined psychologist Ira Noveck to test first- and second-year college students by means of a questionnaire that required them to evaluate some explanations of elementary astronomical facts. For example, the students were asked to judge the truth of certain descriptions of relationships between celestial bodies. Naturally, we had to separate memorized facts from knowledge that required at least a bit of reflection on the position of celestial bodies; we were interested in this latter knowledge.

What Do We Think We Know About the Moon?

The students were reminded that a lunar eclipse is produced when the earth casts its shadow on the moon. Then they were asked whether lunar eclipses are more frequent when the moon is full or when it's in its first or last quarter—when there's only a crescent moon. The majority of students *believed eclipses are more frequent with crescent moons.* The problem is that there can only be an eclipse when the moon is full! (The sun lights the earth, which shades the moon. When the moon is in its first quarter and appears to be only half-lighted, it is illuminated "from the side" in respect to the earth, so the earth cannot shade it.)

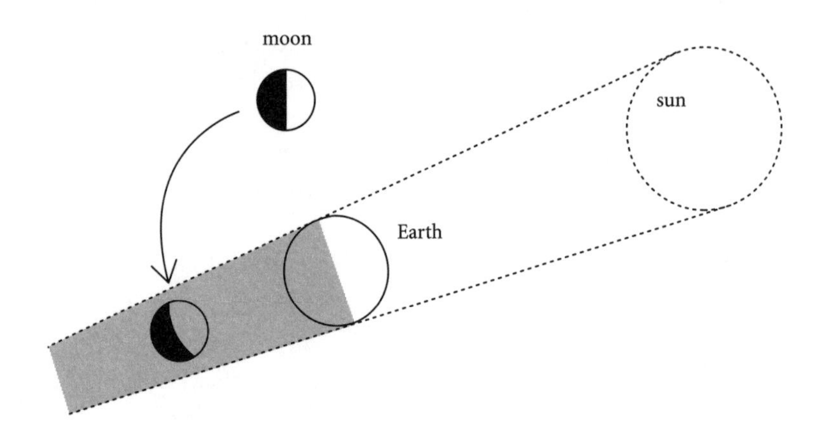

A lunar eclipse is possible only when the moon is full.

A review of the answers to this questionnaire will help us to comprehend the difficulties presented to natural astronomers; in some cases we chose the questions precisely because they bring us back to the key problems of astronomy's prehistory. Some answers to our test are really surprising, particularly where shadows come into it. A significant percentage of the students claim that the *phases* of the moon—for example, the fact that the moon is half-obscured—*are due to the shadow cast by the earth.* This would mean that we were almost always in a situation of partial or total lunar eclipse, with the only exception being the full moon—which, as we have seen, is the only phase in which a lunar eclipse is actually possible! After the test one student (with a background in college mathematics) told us that he was so certain that the phases of the moon were due to the earth's shadow that he no longer wanted to believe the definition of lunar eclipse. He knew the correct definition ("A lunar eclipse is caused by the shadow cast on the moon by the earth"), but suddenly this seemed wrong to him, since it contradicted his idea of the phases.

We might note that by simply observing the sky, without studying astronomy, one could recognize that the phases of the moon have nothing to do with the earth's shadow. Have you ever noticed that the sun and the moon are visible above the horizon simultaneously for almost half the lunar month? The earth is not between them, and so it cannot cast a shadow on the moon (instead, the moon shows a shadowed side, because it's often near its first or last quarter). How could our test students have missed this? The fact is that many students actually claimed that *the sun and the moon never both appear in the sky together.* Doesn't it seem obvious? If the sun reigns in the daytime, the moon reigns at night. One test subject observed: "You've got to agree that it's fairly rare to see the moon during the day. When you do, it's as if the moon showed up early by mistake." One person even joked that the moon is more important than the sun, because it gives us light at night, when everything is dark. There are some American Indian languages in which the sun and the moon have the same name. And it would be interesting to study the skies in paintings up through the Renaissance; at a superficial count I would say that the moon almost never appears in daytime skies, and almost always appears in night skies.

Another classic confusion is the mixing up of the hidden side and the shadowed side of the moon. Half of the moon is illuminated by the sun, and because the moon spins on itself (in respect to the sun's rays), the hidden side of the moon has a collection of phases just like the visible side. Contrary to what many students think, there is no "dark side" of the moon, if by this we mean a lunar hemisphere that is always in the shade. (The geographer Jonathan Raper told me that when he was little he was given a lunar sphere half painted dark. That's a masterpiece for the museum of didactic horrors!)

Natural astronomy thus clashes not only with difficulties of perception but also with the problems raised by human reasoning about objects in space, in particular about objects that rotate around others and require a change in points of view. It is against the background of these difficulties that we should reread—with admiration—the history of the first (often strange) astronomical hypotheses.

Shadow Drawings on the Chalkboard of the Sky

As Parmenides showed us, the moon opens the door to natural astronomy. The moon was the first celestial body to be carefully studied. And shadow really helped the construction of this theory. For example, people considered the phases of the moon, the succession of the moon's *own* shadow—the parts of the moon that take turns in shadow. (Interest in the phases of the moon goes back to ancient times. A bone found in the French Taï cave in 1969—a remnant of the Upper Paleolithic period, from about 10,000 B.C.—shows regular notches in groups of twenty-nine; perhaps the bone's owner was registering the phases from one full moon to the next.) We are interested in lunar phases here because they are fundamental for understanding that the moon is a sphere; there are documents to this effect from ancient Greece.

Eclipses were also pondered: the earth's shadow *cast* on the moon, or the moon's shadow on the earth. Phases and eclipses indicate the spatial relationship among the earth, moon, and sun. (The planets have phases and make eclipses that are not visible to the naked eye; their discovery had to await the invention of the telescope in the seventeenth century.) The earliest astronomical discoveries were these: *the phases of the moon and eclipses are shadow plays that reveal to the naked eye the shape and distance of the moon, the shape of the earth, and the distance of the sun.* You just have to know how to look.

Two theoretic breakthroughs have to be seen in the light of these discoveries. The first breakthrough was the mastering of geometry. The second is the hypothesis that light is a vehicle for this geometry. Light carries with it a trace of shadow. This explains why shadows have had such a crucial role in the evolution of astronomy: When geometric reasoning began to be used to untangle the knots of natural astronomy, shadows became immediate models for perceptive verification. They made sky shapes visible.

Shadows are the traces left by light meeting the bodies that it finds along its way. If you know how to read a shadow, you can reconstruct the history of the encounter. The shadow speaks. It tells us what objects are like, by sketching them with chiaroscuro; and it shows us the position of objects in relation to one another, in relation to the light source, and in relation to the space they're in. Light snaps a picture of objects' alignment: if two objects seem to be lined up (like a target in the sight of a rifle), this means that a ray of light touches both objects before reaching our eyes. (Things are complicated by refraction and by quantum interference, but these are phenomena that come into play at a further level of complexity.) Shadow is useful because it *makes visible an alignment that we could not otherwise see.* Three points lie along the same line: the light source, the obelisk's tip, and the tip of the obelisk's shadow. Of course, as long as we're dealing with obelisks, trees, and—at most—mountains, we can check the alignment by moving so that we ourselves stand at one point along the line of light. But things are different when we turn to astronomical objects we cannot travel to: then it's best to use shadows.

The Moon Shades Itself

An understanding of the phases of the moon creates some problems with perception, and it's no surprise to hear that such problems arose even at the dawn of astronomy. One knotty problem was the question of the moon's shape. Nowadays we know that it is (nearly) spherical; but was that easy to discover?

If we look at the moon in only one phase, any phase, it's not easy to conclude much about its shape. Many languages, including English, say that the moon is "waxing" or "waning," and the reference is to the moon, not to the light striking it. The shadowed part of the face turned toward us is not usually visible. Sometimes when a thin cloud passes over the waxing or waning moon, the crescent can be so luminous that it gives the impression that the moon is passing *in front* of the cloud. Given that the shadowed part isn't visible, the crescent moon seems to be the whole moon.

The moon is luminous like a frosted glass lamp; the light seems to come from inside. The poet and philosopher Xenophanes (fifth century B.C.) described it like this, and said that we see the phases when the lamp is partly extinguished. Parmenides, on the other hand, was convinced (as we have seen) that the light of the moon came from the sun; the moon is not a lamp that gets extinguished, but a body with some parts in the shade.

In these early theories we see the passage from a kind of astronomy that simply registers the regularity of certain phenomena—in order to *predict* appearances—to a more ambitious kind of astronomy that tries to *explain* the reason for appearances. At this point the hypotheses began to multiply. Anaxagoras (496–428 B.C.), introduced by the ancient historian Diogenes Laertius as someone with clear ideas about the meaning of life (we are born "to study the sun, the moon, and the skies") and with equally precise opinions about the structure of the sky, declared that "the moon has a terrestrial nature and has plains and ravines," and that even the spots on the moon are shadows. But

Anaxagoras did not have a convincing explanation for lunar phases: he thought the moon was a *disk* and not a sphere. A disk, as opposed to a sphere, cannot have phases like those of the moon. One can see why the moon might seem to be a disk. The moon spins on itself with a period of rotation that corresponds to its revolution around the earth, and (with slight discrepancies) it always shows the same face to whoever is looking at it. The fact that the period of the moon's spin coincides with the period of revolution has been an obstacle for astronomy. The moon has always seemed static and unchanging. If the moon spun more quickly on itself and thus did not have a hidden face, we could easily see that it's a sphere. It should be added that the moon's particular luminosity is also an obstacle. Its surface seems equally bright from wherever it's viewed, and it doesn't have the shadings characteristic of spherical objects. To understand the shape of our moon we can count only on its shadow line.

Shadow Is Ambiguous, but the Moon Is Not a Bowl

A problem arises here. Seen from a certain distance, on an isolated object, *shadow is ambiguous.* It can indicate either concavity or convexity. Painters have long known this, and psychologists of perception have studied the concave-convex inversion created by shadows. In the top pictures on the next page the circular configuration at right was obtained by reversing the shading of the left-hand configuration. When one of the two figures appears convex, the other one appears concave.

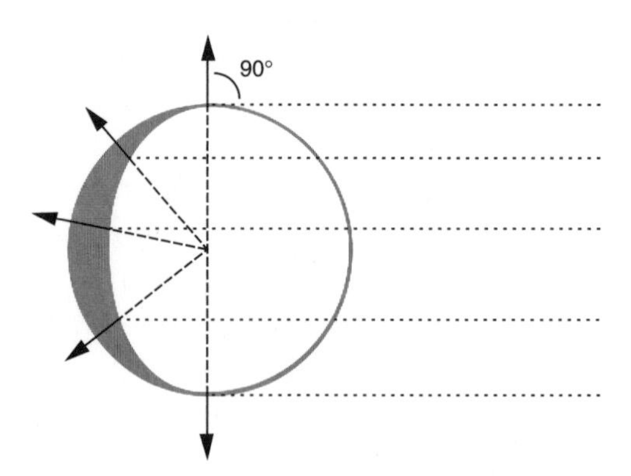

A valuable bit of geometric information: on the moon's shadow line the sun's rays make a right angle with the vertical axis.

Here is the hypothesis: based on what we can see in the sky, could the moon be a bowl whose inside we can see? Though seductive, this hypothesis neglects the sun. (For example, if the left-hand figure in the next picture is a bowl, then the sun must be on the left.) The shadow might suggest that the moon is a bowl, but then the same shadow immediately denies that it is so, because the lighted side is the side closer to the sun.

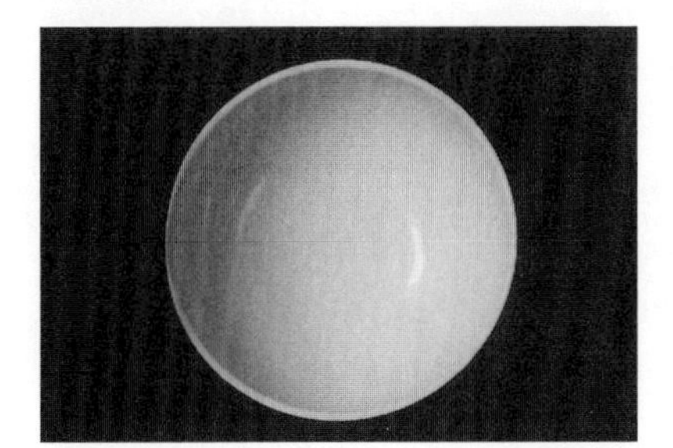

Bowl or protuberance?

For any possibility you can name, there exists a philosopher who turned it into a theory. According to Diogenes Laertius, Heraclitus (550–480 B.C.) declared that celestial bodies are bowls with their concave sides toward us. In this concavity they hold a flame whose light reaches us. The bowl spins on itself, and its edge hides one part of it: these are the phases. (Eclipses happen when the moon's and the sun's bowls are overturned.)

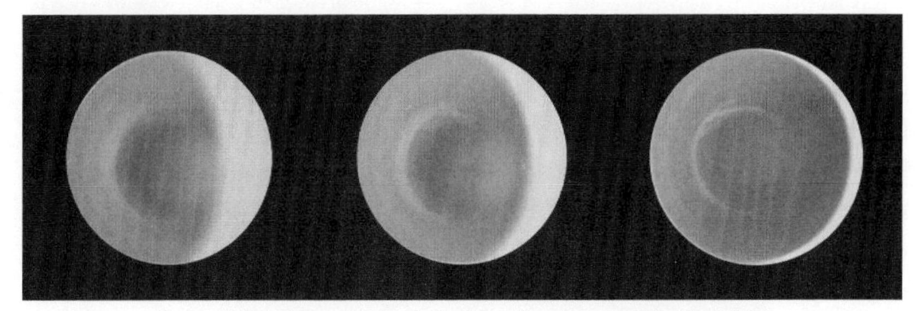

If a bowl like this one (seen here from the concave side, the fillable interior) can have phases, why can't the moon be a bowl? Because its phases are the inverse of a bowl's phases.

These are fine theories, but where do they take us? Not very far, because once again they make us ignore the role of the sun. What happens to the sun's light once it's intercepted by the moon? These hypotheses don't explain that. But if on the other hand we hypothesize that the moon is spherical and that it is lighted by the sun, we can explain lunar phases simply and convincingly.

Let's review the principal problems for a natural astronomer. In the first place, celestial bodies are too far away and too big to estimate (with the naked eye) what might be the relations between their distances, if we don't consider their position in relation to one another. During a solar eclipse, for example, we see the moon pass *in front of* the sun; but at sunset the sun doesn't seem much farther away than the moon. Furthermore, the vault of the day or night sky offers no visual hook to help us determine distances; indeed, it confuses us because it seems closer or farther depending on whether one looks toward the zenith or toward the horizon. And finally, looking at the sky is not enough: we have to consider the figures that shadows reveal, and this becomes even more important when we start to understand the nature of lunar phases.

It's surprising that even with so few elements available, a brilliant astronomer who knew how to use geometry and had a sense of shadow could figure out a good approximation of the distances and size of the sun and moon. His reasoning transformed the sky into a great chalkboard where shadows sketched out a geometric demonstration.

Aristarchus and the Shadow Lines of the Half Moon

That brilliant astronomer was Aristarchus of Samos. The only text of his to have come down to us is his treatise *On the Sizes and Distances of the Sun and Moon.* This is really the only astronomical work that we have from that era. (Aristarchus lived in the third century B.C., and he was quite a character: historians attribute to him one of the first theories placing the sun at the center of the cosmos.) He applied the methods of geometry to an astronomical problem: the astronomer set to work drawing, not just observing and recording. The treatise asserted something stunning that leapfrogged right past perceptive astronomy. It said that there was a big difference between the sun–moon distance and the earth–moon distance: the sun is *nineteen* times farther from the earth than is the moon. Aristarchus's reasoning is perfect in its simplicity. His crucial observation requires scrutiny of the moon's shadow line at a particular point in the month. We have to wait for the day when the line

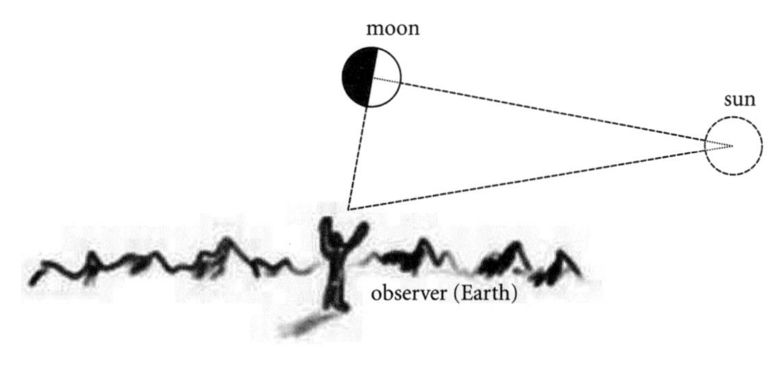

Aristarchus using shadows to draw in the sky.

stretches to divide the moon exactly in half. At this point we must observe the half moon in the daytime, when the sun is also visible.

What we see is an immense triangle. At the center of the moon the line of our gaze would meet a sun ray; because what's visible is a *half moon,* the sun ray and the line of sight form a right angle. At this point all we have to do is measure the angle overhead that indicates the disparity between the positions in the sky of the sun and the moon (this can be done, albeit imprecisely, by pointing one hand at the moon and the other at the sun). Aristarchus measured it as 87°, slightly less than a right angle. This means that the moon-earth-sun triangle has this shape, more or less:

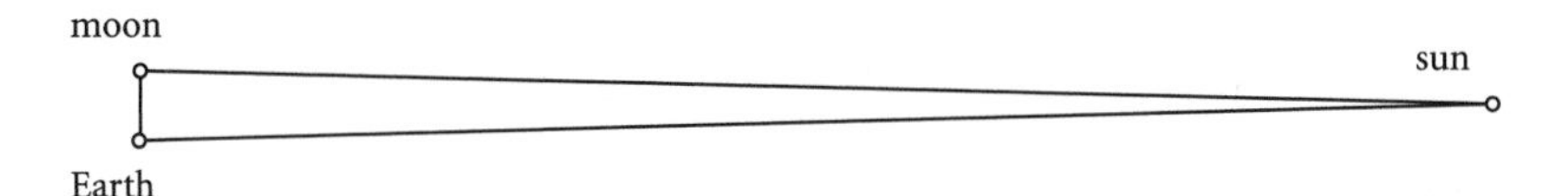

The unexpected proportions of Aristarchus's triangle

So it's extremely long and narrow: the sun is at a very distant apex. With an angle of 87°, the distance between the earth and the sun (the hypotenuse of the triangle) turns out to be approximately nineteen times the distance between the earth and the moon (the minor cathetus). Today we know that the actual angle is closer to a right angle: it measures 89°50', just a sixth of a degree less than 90°. The discrepancy hardly seems important, but with small angles and big triangles we can never be too precise. And in fact the sun lies an average of 93 million miles away from the earth, while the moon lies an average of only 240,000 miles away; the sun is not twenty times but four hundred times farther from the earth than is the moon. So Aristarchus made a fairly large error in locating the sun, as he did, one-twentieth of its real distance away. But it should be noted that if we use his method the error is difficult to correct even with today's observational instruments. The method itself is geometrically impeccable. But it's useless in practice because of the difficulty of catching the exact moment when the moon is swathed half in light and half in shadow. The shadow line is too imprecise. (In any case, twenty centuries passed before people understood that the method was too rough.)

We should not forget that Aristarchus's treatise was most of all a geometry exercise. So however imperfect they may be, these numbers are food for thought. Aristarchus laid down a challenge for us: *this is the geometry of the situation; if you want better results, find a more precise measurement.* And even though he was wrong about the astronomical truth, Aristarchus can be credited with having greatly reduced our perception error. To the naked eye the sun doesn't seem to be nineteen times farther away than the moon, and locating it so far away drastically reshapes our mental map of the sky.

The ancient Greeks' examination of the limits of knowledge was not indifferent to these first extraordinary hypotheses. On one hand, if the hypotheses are correct, then perception tricks us—at least when we look at the sky. On the other hand, what guarantees that the hypotheses are right? The guarantee is that the calculations have been done properly; in other words, that the truth of the facts can be verified (for example, we see that what's rising in the sky is indeed a half moon) and that the proper mathematical tools have been used to arrive at a conclusion. Maybe it would be too much to claim that philosophy was born from the pondering of tricky lunar phases; and two millennia of debates have run aground on the deeply philosophical problem of the distinction between appearances and reality. A far more interesting problem can be discerned in the shadow of celestial bodies: how is it possible to apply mathematics to reality?

VII Eclipses, Shadow Cones, and Pyramids

All shadows whisper of the sun.
—Emanuel Carnevali *(poète maudit)*, 1919

A Message from Space

Literature from all around the world attests to man's dismay at the sight of a solar eclipse. Archilochus, a Greek poet who worked between 680 and 640 B.C., wrote: "Nothing can surprise us anymore and nothing seems impossible now that Zeus, father of the Olympians, has changed day into night. We can now believe anything, expect anything. Don't be surprised if in the future the animals of the earth switch places with the dolphins and the dolphins prefer the mountaintops." The Old Testament indicated that an eclipse would be one of the signs of the day of wrath: "The sun shall be turned into darkness and the moon into blood." To make the death of Christ even grimmer, the Gospels describe a darkness falling on the earth that recalls the darkness of an eclipse. The Takana (in Bolivia) speak of the "Death of the Sun" that causes animals and objects to revolt against humankind. Eclipses have been held responsible for provoking bloody wars, ruining dynasties, and triggering earthquakes and floods.

I have always been suspicious of these tales of terror. I had seen hundreds of photographs and films of eclipses, I had witnessed two partial eclipses, and I had never found anything particularly worrisome in the picture of a disk hiding another disk. Maybe writers in the past were just playing the easy emotional card? But the fact that so many accounts agree on the dramatic nature of eclipses started me thinking. After seeking some ulterior justification (after all, I can't write a book about shadows and ignore the most majestic shadow cone of all), I set off to see the eclipse of August 11, 1999, on the Black Sea; forecasters said this spot would offer the best combination of clear skies and a long period of totality. With some skepticism I awaited the conjunction in the company of a wide array of astronomers and astrophiles. Convinced by now of the banality of the whole thing, I watched the partial phase, I recorded it with a few drawings, and at the last minute I took off my protective glasses while the moon finally shifted over to cover the sun.

And at that point my mind was changed.

The two or three things that I expected to see in the sky were just a minor part of the spectacle. A total eclipse is by far the most impressive natural phenomenon that we terrestrials can witness. The staging doesn't lessen its brutal effects. The temperature drops. A mysterious cold wind starts blowing. The shadow comes running up like a hurricane on the sea. The light collapses, and in just a few seconds a metallic night falls—it comes on so fast the mind is not ready for it. On the horizon, unreachably far away, are the vestiges of daytime: an orangy twilight all around, as if a set designer made a mistake in projecting a sunset. In the midst of all this is a sun that's no longer a furnace but just an unlucky rock: its shining fringe is like the silver mane of hair of some aged celestial divinity; and stars glitter again, caught out of place in this out-of-joint nighttime.

If I had to draw an analogy, I would say it's like being transported to an unknown planet. The brain reacts spasmodically, unable to find any resources in the catalogue of memory: it can't think of any way out. Disheartened after a series of attempts, it suggests: dream landscape? Discomfort is inevitable: you just want to get it over with quickly.

Then you get addicted. I had a relapse in 2001: I flew to Zambia, where, on the day of the summer solstice, I was faced with an even more majestic scenario. Before the eclipse, invisible clouds were lining up along the horizon, too bright to be seen. They became visible as the shadow enveloped them, popping out of nowhere, line after line of clouds coming closer and closer, as if they were being created on the spot right before my stunned eyes.

Spending a few minutes in the shadow of the moon sets all our odometers back to zero. It brings us back to that period in our history when we realized that we were part of a system far more majestic than that of the simple things of daily life. It allows us to imagine the depth of the astonishment that our species felt when it began to launch itself mentally into the space beyond Earth.

The Need to Explain Eclipses

In Aristarchus's time the basic mechanism of eclipses was understood. But there had been obstacles in gaining this understanding. Here too there were problems with the limits of perception. Through observation alone it's hard to figure out the reason for a solar eclipse (that the moon gets in between the sun and the earth). The sky is often cloudy, and when it is, only a profound darkness is visible during the eclipse. When observing an eclipse with the naked eye, you cannot see the moon approaching the sun (because the atmospheric luminosity is too strong). It's surprising how blinding even a tiny part of the sun's surface is: an eclipse is not noticeable until the moon has eaten through three-quarters of it. When you notice the eclipse, you can see only the edge of the moon's silhouette, and to recognize the moon you have to know that, at that moment, only the moon can be located in that part of the sky. So there are a lot of difficulties, so many difficulties that certain cultures don't give the moon any role in solar eclipses. The cosmology of ancient China invented a dragon that sometimes bit into and sometimes swallowed the sun (the Chinese word for "eclipse" was "bite"). In Arab astrology a phantom "eighth planet" was thought to be responsible for eclipses. Western iconography documents such hypotheses up until relatively recently, in some cases mixing them up, as in the drawing on page 72.

It is equally hard to understand that a lunar eclipse is caused by the projected shadow of the earth. Here the only thing to be seen is that the moon slowly goes darker; in some cases it changes color, growing reddish.

The correct explanation for lunar eclipses seems to go back to Babylonian astronomers of the first millennium B.C.; in the Greek world the explanation has been ascribed to Thales of Miletus (624–548 B.C.). Records of the history of Greek astronomy are hard to assess; but they do testify to the appeal of shadows. The following stories are told: Anaximander (610–547 B.C.) thought that the moon was like the wheel of a chariot, and that eclipses depended on the turning of the wheel. Anaximenes of Miletus (585–520 B.C.) considered lunar eclipses to be a phenomenon of shadow—but he was wrong about the source of the

In ancient times people believed that a dragon bit into the sun during an eclipse. In this illustration from a 1497 almanac, the dragon simply hides it.

shadow. For he believed that there were solid bodies—actual secret planets—that intercept the sun's rays and cause eclipses (as well as the phases of the moon). Some ancient authors thought that Anaxagoras was the first to offer a correct explanation of eclipses—that the earth casts its shadow on the moon. (Anaxagoras asserted also that the sun was an incandescent rock much larger than the entire Peloponnesus, a heretical theory for which he risked death; he was the first in a series of researchers who have been persecuted for religious reasons by political authorities.)

Anaxagoras was so pleased with his interpretation of eclipses that he began to see shadows everywhere. For example, he maintained that the Milky Way was visible because of the shadow cast by the earth. The idea was that the sun "snuffs out" the stars beyond the earth's shadow, and that only the stars in the shadow glitter, tracing the Milky Way in the sky. A shadow can reveal light that would otherwise be invisible. Again, we see that there's no bizarre theory that hasn't been promoted by some philosopher; but the fact that Anaxagoras defended this particular theory seems to me to confirm the intellectual pull of shadows. Anaxagoras understood that the earth is a body that casts its shadow in the sky, and he gave free rein to his hypotheses: the shadow of the earth is somewhere in the sky even when there's no moon to make it visible. (As the historian Thomas Heath observed, if the Milky Way is the only part of the sky in shadow, the moon should be eclipsed every time it passes by the background of the Milky Way—which doesn't actually happen. In this case we could almost say that the *absence* of a shadow teaches us how things work.)

Let's try to re-create the audacity and perhaps the sense of wonder of these sky intellectuals at the moment when they concluded that a lunar eclipse is caused by the shadow of the earth that intercepts the light of the sun. When a lunar eclipse is explained as the result of a projection of the earth's shadow in the light of the sun, these three heavenly bodies are linked in a single system. There is no metaphysical gap between terrestrial and celestial things—on one hand a vast, heavy, dark land, and on the other the airy lights of the day and the night—there is only distance across space. Denuded of light, the moon reveals itself for what it is: a cold, inert rock suspended in the sky. Which means not only that the moon and the sun are like the earth, but also that the earth is like the moon and the sun: the earth *is a body that hangs somewhere in space.* The metaphysical gap is bridged by geometry. And the dimensions of the bodies also demand that everything that was once known about distances be rethought using new concepts. The earth's shadow cast on the moon gives us an idea of the shape and size of the earth. The first known demonstration of the fact that the earth is spherical and that it's bigger than the moon goes back to Aristotle (384–322 B.C.). If the earth were not spherical, "lunar eclipses wouldn't show the lines that we see. . . . In an eclipse the edge line is always curved," and only if the earth is spherical can it cast onto a spherical moon a shadow line that's a segment of a circle.

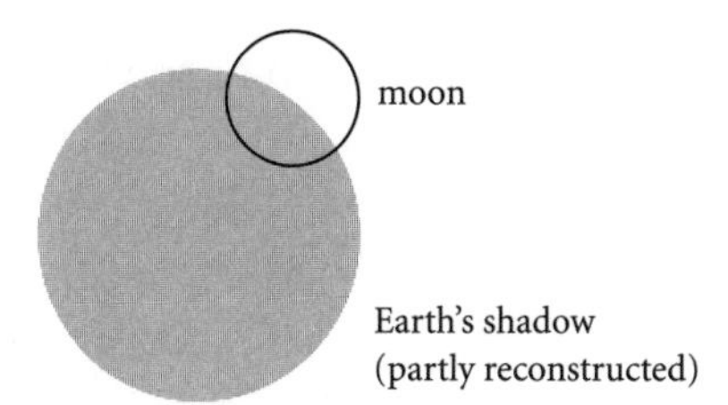

A lunar eclipse shows that Earth is much larger than the moon (if we know that the sun is very far away), and also that it's spherical.

If the earth were a cube, it couldn't make the shadow it makes on the moon. And also, given the great distance of the sun, if the earth were smaller than the moon, it couldn't swallow up the moon with its shadow.

The Legend of Hsi and Ho and the Difficulty of Forecasting an Eclipse

The story of Hsi and Ho, astronomers in ancient China, is a sad tale. Among other jobs, they had the task of predicting eclipses. The complicated calculations necessary were not made any easier by the two men's propensity for tippling. It was inevitable that the unforgivable would happen: an eclipse occurred that they had not predicted (or perhaps they forgot to perform the necessary rites of exorcism). The emperor sent for them, they were found drunk, and they were decapitated.

Of all cosmic conjunctions, eclipses are the most troubling; even cold, rational minds find it hard to resist their appeal. (I indulged in a slight chill when I discovered that I had been conceived around the time of the solar eclipse of February 15, 1961: was I perhaps always destined to study shadows?) Considering the effects of eclipses, it would seem natural to worry about what's happening and to see whether it can be fixed—the sun and the moon are pretty valuable. Some cultures call for rites or games to free them. People shout at the ailing sun or moon: "Get well!"

It would be even better to predict the return of the shadow, in order to take adequate countermeasures (if there are any). But what can be done if one hasn't got an astronomical theory? One has to exploit the strange fact that eclipses are (somewhat) cyclical. The repetition of eclipses indicates that the cosmos conceals deep and complex regularities; the ability to predict an eclipse is a sign of awareness of the hidden rhythms of the universe. The documentation shows reliable data such as the Babylonian astronomers' recognition of recurring eclipses (starting in the eighth century B.C.); but often this data is embellished with legends (such as the tale of Hsi and Ho).

The prediction of eclipses is a delicate matter. In the first place, you have to discover that a lunar eclipse is possible only during the full moon, and that a solar eclipse is possible only during the new moon. We saw that many students in our test regarding naïve astronomical concepts didn't clearly see the geometric relationship between the full moon and the lunar eclipse, so understanding this relationship is an interesting intellectual achievement. But this achievement is not enough. We don't get a lunar eclipse with every full moon (and we don't get a solar eclipse with every new moon). The great leap forward was made by the Babylonian astronomers of the first millennium B.C. who, across several generations, compiled a systematic catalogue of lunar observations and eclipses. Using the catalogue, they were able to single out some recurrences and to discover that eclipses are cyclical and that it is therefore possible to predict them. This result is also stunning, considering that partial lunar and solar eclipses easily go unobserved. It is not at all simple to keep a list of eclipses (and if the catalogue were incomplete it would be hard to grasp that recurrences happen).

The House of Celestial Appointments

Why are eclipses cyclical? Why, after a certain amount of time, does a certain type of eclipse reappear? It seems strange, but the fact is that eclipses could be much more cyclical. If the moon revolved around the earth on the same plane as the earth's revolutions around the sun, we would have a solar eclipse with every new moon. But the moon's orbit is on a plane tilted five degrees off the orbital plane of the earth (so that when it's new or full, it cannot be aligned with the earth and the sun; it's a bit "higher" or "lower"). And furthermore, the abstract points at which the orbits intersect (the nodes, in astronomical terms) move slowly in the direction opposite to that of the moon's revolution. This means that while twenty-nine and a half days (one lunation) pass between two moons, the moon itself takes less time—twenty-seven and a quarter days—to move from one node to another (this is known as a "draconian" month, named for the dragon lurking at the nodes).

The whole thing is now but a series of appointments in the cosmic datebook.

On Monday you meet with Helios and Diana. If Helios visits you every two days and Diana every three days, how long will it be before all three of you are together again? Answer: six days later. All you have to do is multiply. Let's say we're in a favorable situation for a lunar eclipse, when the full moon meets a node. To come back and meet the same node when the moon is full again, we do the multiplication and see that 223 lunations have to pass, or 242 draconian months: 18 years and 10 days. (This all works well for a few millennia; if we go too far beyond that, the interactions among the earth, the sun, and the moon make the cycles a bit chaotic.) In all this time the sun is on the opposite node nineteen times. If in the initial situation there was in fact a lunar eclipse, there will be another one 18 years and 10 days later. The interesting thing is that in this period the moon will have made, more or less, 239 orbits (the moon takes 27½ days to complete an orbit, an anomalistic month in astronomical terminology). And it will have come back, more or less, to the same position in its orbit, and be at the same distance from the earth, as it was at the start of the cycle. Which means that not only are eclipses cyclical, but the types of eclipse are too, fairly regularly. If the cycle begins with a total eclipse, after 18 years and 10 days it will start again with a total eclipse in almost identical conditions. Many of these numbers were known to the Babylonians around the fifth century B.C. Among the Greeks it appears to have been Thales who grasped the science of shadows.

The Legend of Thales and the Inauspicious Influence of Eclipses

The historian Herodotus (in the fifth century B.C.) wrote that after many years of war between the Lydians and the Medians, with the advantage going back and forth, a battle was interrupted by a solar eclipse that had been predicted by Thales; the phenomenon convinced the opponents to put down their arms and make peace.

There is an important difference between the documents of Babylonian astronomy (tablets with predictions and records of regular recurrences) and those of Greek astronomy (literary texts). Quantitative methods can be applied fairly directly to the former; for the latter we must be satisfied with pretty stories that often have no verifiable confirmation. Such is the oft-told story of Thales' prediction of a solar eclipse. It's a problem not only in Thales' case. In a solar eclipse the point of the moon's shadow cone follows tracks that change from eclipse to eclipse; there is no cycle of solar eclipses for a certain location on the earth (for Babylon or Athens, for example). So you cannot predict a solar eclipse for a given spot based exclusively on the list of earlier solar eclipses. On the basis of the list one can predict only a favorable moment for an eclipse, or when an eclipse cannot happen. This makes it pure fantasy to attribute to Thales the prediction of a solar eclipse, or to attribute such a thing to any ancient astronomers, astrologers, or diviners.

With our present-day knowledge of the solar system, we can now calculate future eclipses by studying the movements of celestial bodies.

Furthermore, nothing keeps us from calculating past ones: all we have to do is turn back the clock. In 1887 Theodor von Oppolzer published his monumental *Canon of Eclipses,* listing 8,000 solar eclipses and 5,200 partial and total lunar eclipses from 1207 B.C. to A.D. 2161. What is the interest of this series of calculations (done by hand)? Eclipses are fixed appointments that intersect with human history. In theory, they should help the historians. Because we can go back and determine solar and lunar eclipses, it's easy for us to date the events of the past by using eclipses as a great calendar (which is also disdainfully indifferent to arbitrary human calendars). Could this save the story of Herodotus? Even if Thales' eclipse could not have been predicted by the philosopher, perhaps its date, inscribed in the sky's order, can allow us to figure out the date of the battle between the Lydians and the Medians. (Consulting Oppolzer's *Canon,* some modern historians have fixed it at May 28, 585 B.C.)

In real life, things are harder, for reasons independent of astronomy. It sometimes happens that ancient tales mention an eclipse only because it's a spectacular event, even though it didn't happen at all! Certain events *ought to* be accompanied by great cosmic signs: big battles, deaths of kings and queens, or the birth of evil heirs who will bring their families to ruin. If there actually was an eclipse just before or just after the Great Event, then the date of the eclipse will be brought nearer to the date that the narrator cares about. And if there is no eclipse in the neighborhood at all, one will be invented from whole cloth. Christ was crucified at Easter time, so it was during the full moon: it's impossible for there to have been the solar eclipse that is described in the Gospels. The closest solar eclipse happened on November 24, in A.D. 29, and it was not visible from Jerusalem. A chronicle of the deeds of the bishops of Liège speaks of how Heraclius (who died in 971) managed to placate soldiers who were terrified by a solar eclipse during a battle in Calabria. That eclipse must have been the one on December 22, 968. But according to the chronicle, this eclipse was also supposed to have happened during the full moon, so we can only conclude that the chronicle is not completely reliable.

We also know that the business of the wise man reassuring the army when the sun disappears or the moon is dimmed is a literary commonplace with illustrious antecedents in Pliny and Titus Livy. Quintilian even recommends that orators get acquainted with the whys and wherefores of eclipses, because you never know, you might find yourself having to explain it to soldiers right in the middle of a battle (apparently this happened to Pericles and to many other generals). The one example that no one wants to emulate is that of Nicias, commander of the Athenian armies in the siege of Syracuse in 413 B.C. A lunar eclipse terrified him and made him delay his departure by a month; his enemies got reorganized, then fought the Athenians and decimated their fleet. For being ignorant about the workings of the universe, Nicias went down in legend as a victim of superstition.

Shadow Triangles and the Mystery of Thales

Shadows measure out the cosmos, but the association between shadow and measurement was part of the Greek world even apart from geography and astronomy. Thales turns up here too: "Hieronymous declares that he measured the pyramids based on their shadows, after having observed the moment when our shadows are equal to our own heights."

Thus wrote the historian Diogenes Laertius. It's a nice story. Thales went to Egypt, picked up the basics of geometry just by rubbing shoulders with the priests there, and immediately showed his genius by using a simple and brilliant method to solve a problem that had apparently stumped his masters and their masters since time immemorial, even though they had the pyramids right in front of their noses.

It's a nice anecdote, but it's fishy.

Thales is a curious figure. He is introduced respectfully as one of the seven wise men, and at the same time he is criticized for a certain tendency toward abstraction that made him tumble into a ditch, for example, while walking along looking up at the stars. In the stories, Thales retaliates with a clever speculative move. He predicts a good olive harvest, and he rents some olive mills, which he then sublets out at high prices. This move was made only in order to show how skillful he is, and how clever are philosophers in general. In all his generous and slightly snobbish messiness, this character cannot help being appealing; he wrote only two works, *On the Solstice* and *On the Equinox*, "declaring that any other topics are too elusive."

Thales was one of the first cases of the personalization of science. Some outstanding mathematical discoveries are attributed to him: they say he demonstrated that any angle inscribed in a half circle is a right angle; that the base angles of an isosceles triangle are equal; that opposite angles are equal; and the diameter of a circle divides the circumference in two halves. (You would be proud to be the first one to discover even one of these things.) But these are geometric propositions too fundamental to have gone unobserved before Thales' time.

Shadows and Geometry

Thales went down in history for a theorem that bears his name. Imagine the shadow of a post in the ground. Try to imagine the whole area that's without light, the zone hanging in air between the post and the projection plane (the ground where the shadow falls). This zone is a triangle: one side is the post, another is the shadow cast on the ground, and the third—invisible—side links the tip of the post to the tip of the shadow. The properties of this whole shadow zone are the properties of a triangle. Now, the longer the object casting the shadow grows, the longer its shadow—that is, the longer will be the ground side of the whole shadow zone. How much longer? If the post doubles in length, the shadow will double. Only if, naturally, the light source doesn't move and the projection plane doesn't tilt. We know that people ruminated over this in ancient times because of the anecdote about Thales and the measurement of the pyramids.

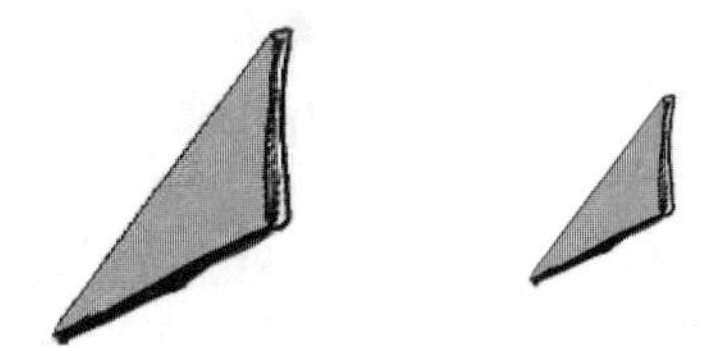

Shadow as triangle: the birth of geometry.

The narrative around the tidy story of Thales in Egypt is complex. Thales is associated with Egypt just a few lines earlier in Diogenes' text, which quotes a source who "says that Thales, having learned geometry from the Egyptians, was the first to inscribe a right triangle in a circle; then he sacrificed a bull." So we learn that the Egyptians had only rudimentary geometry; so much so that Thales could advance it by such steps as the inscription of a right triangle in a (semi)circle, an achievement that certainly would merit a bloody celebration like his. The idea that the Egyptians might have reached an advanced stage of geometry (tradition says that they were brought to it by the need to mark the boundaries of their fields over and over again after the Nile's floods) doesn't seem to be confirmed historically. But this doesn't authorize us to think that Thales' contemporaries in Egypt lacked even the elementary geometry that would allow them to measure the height of the pyramids—pyramids that their ancestors had built with such precision. And the measuring of the pyramids, which doesn't seem to be a very ambitious goal, is a task that should raise our suspicions. It's as if some reporter found it necessary to mention a great practical success alongside Thales' abstract results.

The anecdote about the measurement of the pyramids serves as a reminder that research into shadows, together with a general theorem about triangles, leads us to discover the properties of objects that are otherwise too large or too far away. The tip of a building is hard to reach, but its shadow stretches along the ground in front of us, and all we have to do is measure it out with our feet; with Thales' theorem we will immediately know how tall it is.

So the anecdote serves to illustrate the practical bearing of the theorem. The pyramids, once the star of the show, are demoted to a walk-on part.

Forty-five Degrees in the Shade

But we do need to think about the pyramids. Indeed, from a certain point of view the problem is interesting precisely *because* pyramids are involved. An obelisk or a parallelepiped building doesn't hide its height line; this line is simply a physical characteristic of the building, a corner that can be seen and measured. On the other hand, a pyramid masks its height within a pile of stones. So it's a challenge to choose to measure a pyramid, of all possible buildings.

Now, a pyramid hides not only its height but also its base and, along with it, part of the shadow that Thales would need! The shadow cannot be measured directly, and so one has to conjecture and calculate the hidden portion. First of all, we have to see what pyramid is being measured. The Red Pyramid of Snefru at Dahshur has a slope of 43°. In order for the shadow of the tip of the pyramid to reach out and match the real height, the sun has to be at 45° in the sky. But when the sun is at 45°, the shadow of the tip falls inside the pyramid! The pyramid doesn't cast any visible shape on the ground. Let's go to Giza, where the pyramids are steeper. When the sun is at 45° in the sky, the piece of shadow sticking out from the pyramid's base is not very useful: we still have to calculate the length of the hidden part of the shadow.

These circumstances ruin the example. Thales' method aims to be intuitive and easy to apply—because the rules of myth dictate it. That's required of any brilliant solution offered by a stranger to the poor native sages who for centuries had been unable to untangle the mystery of the pyramids.

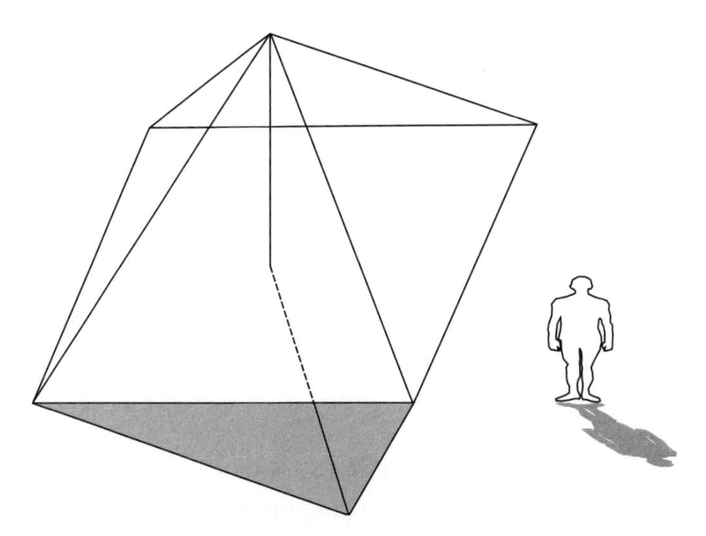

Thales' theorem: when Thales' shadow is as long as he is tall, then the shadow of the pyramid is identical to the height of the pyramid. But how to measure the shadow of the pyramid?

So let's imagine Thales on that famous afternoon when a curious audience waited for the philosopher's shadow to be as long as he was tall. Finally the eagerly awaited moment arrives. At this point Thales runs to measure the margin of shadow spilling over the edge of the pyramid. If all the conditions are favorable, Thales will get a length measurement that he can offer the audience as the height of the pyramid only after he has executed a series of calculations (not too complex but certainly dull). But calculations like this, someone will point out, would allow him to figure out the pyramid's height without waiting for, and then measuring, the shadow: all he has to do is measure the base angle of the pyramid and half the length of one side. So Thales' proceedings won't have made so much of an impression on his audience that they would go down in history. The paradox is that what you need to know already in order to measure the pyramids' heights using the shadow method is enough to measure them without using the shadow method.

If we stick to a prudent declaration and say that Thales was the inventor of the trigonomic method, based on similar angles, for measuring the height of other types of less-pyramidal buildings—particularly obelisks and buildings with vertical corner lines—we can scrub this tale clean of its mythical embellishments. Pyramids grab our attention because they are mysterious in themselves, and already in Thales' time they were two thousand years old; they bestrode civilizations; they were considered to be wonders of the world. It must have seemed natural to insert them into the narrative tradition, in place of some insignificant structure that appeared in an earlier version of the story. But the pyramids weigh down the story with their great mass; their mathematical structure, which strangles their shadows, reveals the hand of an imaginative narrator.

We are left with the metaphysical puzzle of the pyramid that encloses part of its shadow. If we take for granted that only a fragment of the shadow we wish to measure sticks out from the perimeter of the pyramid, we implicitly accept that the other part of that shadow stretches along the inside of the structure. Is the interior of a pyramid shady? Not the inner rooms, but the actual solid interior, built of weighty blocks packed tightly together. Bringing the internal and external portions of the shadow together in one single mental gesture is a magnificent abstraction that reduces shadow to an incorporeal form but raises unsettling new questions.

Shadow and the Distance from the Earth to the Sun

This digression into Thales' shadow triangles permits us to return to eclipses and Aristarchus, armed now with the tools of geometry.

Aristarchus understood the relation between the earth-to-sun distance and the earth-to-moon distance. But all the measures he used in his discussion are relative. The reasoning would work even if the sun were only 12 inches in diameter (as Heraclitus seemed to believe it was): in this case the moon would have a diameter of a bit more than half an inch (and we would be able to reach up and grab it, which is the proof that Heraclitus was surely wrong). Aristarchus then wanted to measure the space between us and the stars and determine once and for all how big are the moon and the sun. Now shadows come into it again. Not the shadows stuck to objects—like the moon's own shadow with its rhythmic phases—but the shadows that move through space casting the image of the bodies that create them. Aristarchus's second shadow method is born of an argument about solar and lunar eclipses. He uses not the shadows that stick obediently to the moon, but only the shadows that sow fear on Earth.

To start with, Aristarchus made a banal observation. In a solar eclipse, by lucky coincidence, the great star is hidden almost completely by the moon. The moon's shadow cone narrows so far that it pricks the earth with its point.

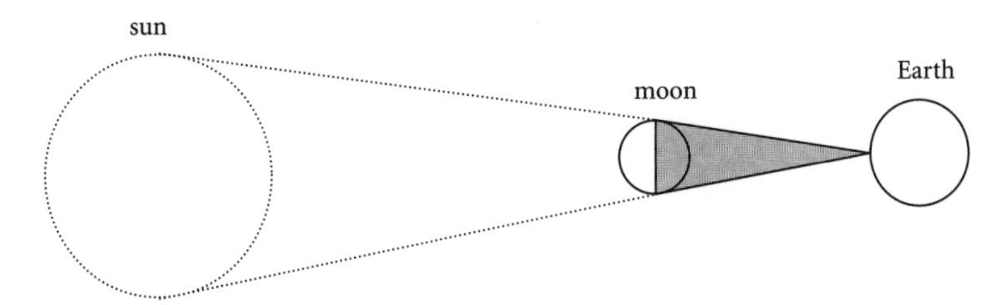

The moon masks the sun almost perfectly; in other words, the moon's shadow cone pokes the earth.

Aristarchus calculated that the sun is nineteen times farther away than the moon and thus worked out that it is nineteen times bigger (this is yet another application of Thales' theorem, which is starting to prove useful). What's missing from Aristarchus's cosmic diagram is a number, a hook on which to hang the construction and transform the relation between the distances into a measurement. But how can such great distances be measured? After using the shadow on the moon itself, and the shadow that the moon casts on the earth, Aristarchus looked to the earth's shadow during a lunar eclipse.

Aristarchus tried this hypothesis: when the earth's shadow cone meets the moon, the cone has narrowed so far that its diameter is twice that of the moon. Following this reasoning about similar triangles (and still applying Thales' theorem) he concluded that the diameter of the sun is six or seven times greater than that of the earth. Judging from the apparent sizes of the sun and the moon (in his treatise Aristarchus used a value of 2 degrees, but Archimedes reported that Aristarchus found a more accurate value of half a degree), one can establish that the moon's distance is 20 (or 80) terrestrial radii from us and the sun is 380 (or more than 1,500) terrestrial radii away from us. The distances of the sun and the moon are grounded in the size of the earth. As in the case of the distance relationships among the earth, the moon, and the sun, these results too are much smaller than reality (the sun is more than 20,000 terrestrial radii away from us), but they are a step in the right direction. Without pondering shadows, there is no way to imagine that the sun could be so far from the earth, nor that it is so big.

How to Use an Eclipse

We can measure the universe by pondering eclipses. But there are other, more immediate uses for an eclipse. I have a sordid story to tell here, so I hope you'll forgive the dubious taste of the illustration.

The advantages of having a good astronomical almanac. (But note the erroneous shape of the eclipse's shadow line: it should be much less curved.) From Camille Flammarion's masterpiece of popular science, Astronomie Populaire.

It was March 10, 1504, and Christopher Columbus (after consulting an astronomical almanac) frightened the Jamaicans by announcing a lunar eclipse that then happened as predicted. At the threshold of modern times, mastery of eclipses served only to indicate the discrepancy between cultures. By this time a shadow line divided those who didn't know about the mechanisms of time from those who relied on those mechanisms to demonstrate their own dominion of space. But time and space, as we shall see, continue to meet at the tip of a shadow.

VIII *The Stolen Sundial*

I mark only the happy hours.
—Inscription on an old sundial

Almost with embarrassment, Pliny the Elder (A.D. 24–79) quickly wrapped up his brief history of Rome's timekeeping and changed the subject: "For a long time daylight wasn't divided adequately for the people of Rome. Let's move on now to the rest of the animals, starting with the land animals."

Pliny deplored the fact that the ancient Romans, those sorry creatures, couldn't measure time based on the sun because they were not interested in the theory of astronomic time. This theory links Earth and the sky in a single great scheme that makes it possible to keep track of the apparent movements of the stars. The appearance of these movements depends on the observer's position on the earth. The earth is a globe, and differences in the observer's position correspond with different heights above the horizon for the stars. (For example, the Pole Star is at the zenith, that is, vertically above an observer at the North Pole; at the equator, it's at the horizon.) If we measure time by the stars, then we should not forget about geography. The Romans didn't understand this; good as they were at sacking cities, Pliny seems to say, they were unable to use their war loot. He tells an edifying tale about this: "Marcus Varro notes that the first public solar clock was brought [to Rome] from Sicily and was erected on a column behind the [orator's platform] during the first Punic War, after Catania in Sicily was conquered by the Consul

M. Valerius Messala . . . in Roman year 491 [264 B.C.]. The clock's lines didn't agree with the hours, and yet people continued to follow the lines for 99 years! They did so until Quintus Marcus Philippus, censor with Lucius Paulus, juxtaposed it with a more carefully conceived solar clock; this was among the most welcome of the censor's works."

What mistake did the sundial thieves make? Because the earth is curved and the stars are different heights above the horizon, solar clocks are calibrated for a specific latitude; if you bring them farther north or south, the lines on their faces no longer correspond with the hours of the day. It's stupid to steal a sundial unless you keep it at the same latitude; Rome, unfortunately, is north of Catania.

OK, so Rome is north of Catania—but not very far (only about 450 kilometers, 4° of latitude). The error that Pliny pointed out is not so bad. Sharon Gibbs, in a study of ancient solar clocks, has calculated that the greatest discrepancy would regard not the hours of the day but the periods of the year: for example, the point on the sundial corresponding to the winter solstice would be touched a month earlier than necessary by the autumn shadow that stretches too long, and would be touched again (when the shadow begins shrinking) a month later than it should be. But within a single day the out-of-place Catanian clock would have been wrong by no more than forty minutes. It's interesting that, faced with such an unimportant error in the practical life of ancient Rome, Pliny derided his ancestors. He wanted to display a more pressing need for precision than what was necessary for daily business, a need that by his time had been largely filled: toward the end of the first century B.C. the Roman architect Marcus Vitruvius listed a good thirteen types of solar clocks from Greece, Asia Minor, and Italy.

Pliny's story brings our attention to the difficulty of setting a solar clock; in the end, the problem is understanding the relative movements of the sun and the earth.

Let's say right off that there's more to a solar clock than meets the eye. The face and the stylus are not the only components. When there is no shadow, the clock cannot function. But since the shadow depends on the sun and since the shadow's movement depends on the rotation of the earth, the sundial is an immense *system* that includes the sun, the earth, and the space between the two celestial bodies. The sun is one piece of the solar clock, and the earth is its engine, just as the spring or the counterweight is the engine of a mechanical clock. If you open up a wristwatch, under the face you'll find balance wheels and gears. If you open up a sundial, you'll find a planet and its star.

The Sky in a Bowl

Everyone has seen a solar clock at least once, in a garden or on the wall of some town hall or church. A stylus, usually parallel to the axis of the earth, casts a shadow that marks the hours on a graduated plane. In the ancient world of Greece and Rome, solar clocks were completely different. They were bowls.

A bowl that holds the sky within it is also among the most ancient of astronomical instruments. If we look for evidence about its invention, the texts bring us back to Aristarchus. (It's really an elementary object; it had probably already long been in use by Aristarchus's time.) The bowl, also known as a *scaphe,* is a hemisphere. A stylus is glued to the center of the bottom (it is also called a *gnomon*—literally, the knower, or indicator); this stick rises vertically as high as the edge of the bowl.

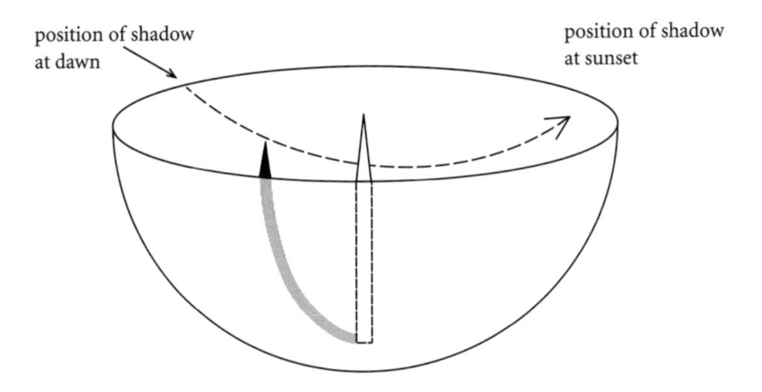

The passage of the sun in the course of the day is traced in Aristarchus's bowl (scaphe).

If the bowl is held horizontally at dawn, the stylus casts a shadow whose tip kisses the edge (on the west, because the sun is in the east). At midday the tip of the shadow is at its lowest point, and it then reaches closer to the border (on the east) at sunset after having crossed the surface of the bowl. If you think about it, you'll see that *the bowl is a reversed image of the sky,* in which the shadow of the point represents the sun.

The higher the sun rises, the lower the shadow of the point drops, and vice versa.

Every point on the celestial vault is linked with a point on the bowl by a line that passes through the tip of the stylus. Turning the tables, we could see the sky as an immense bowl flipped upside down.

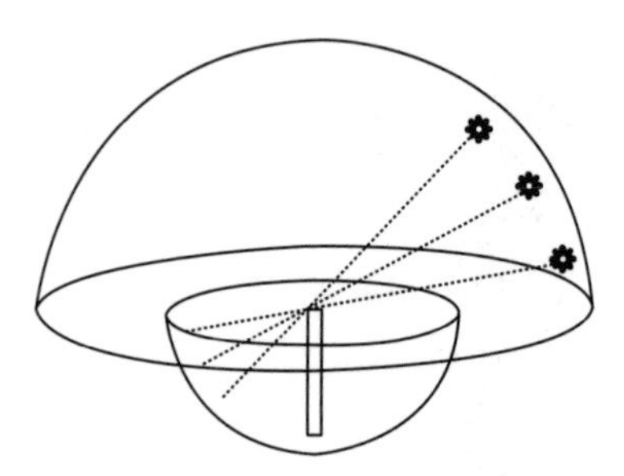

Aristarchus's bowl is an overturned sky—or maybe the sky is an overturned bowl?

By tracing the gradations on the bowl and observing the shadow of the stylus, we can determine the height of the sun and the moon, the only bodies bright enough to cast shadows. (A medieval Arab astronomer said that he saw the faint shadow of the light of Venus; I long thought this wasn't possible, but then in one pitch-dark African night with the new moon, I was able to discern this shadow.) On the longest day of the year the tip of the stylus traces the farthest point reached by the sun in the summer sky; on the shortest day, it traces the point reached in the winter sky.

Unfortunately, we have no surviving examples of the ancient astronomical bowl; its spirit lives on in solar clocks.

What Time Does Your Shadow Say?

Greek and Roman astronomers and mathematicians divided up the day into equal parts, so that daytime hours were longer in the summer and briefer in the winter. In the bowl these divisions correspond to an array of lines that break up the sun's passage and fan out at the bottom, where the interval crossed by the shadow grows longer as the sun passes higher above. But this way of dividing time did not spread beyond a small circle of specialists. In ancient Greece and Rome people didn't care much about the actual time; they made appointments based on the fixed length of a person's shadow, not based on the shadow's length in relation to the person's own height: "We'll meet this afternoon when our shadows are six feet long." This certainly led to some misunderstandings: a short person and a tall person could end up never meeting. If a person is six feet tall, his shadow is six feet long at a certain point in the afternoon. But if the other person is four feet tall, her shadow stretches to six feet much later in the afternoon, when the sun is going down. The real moochers cunningly took advantage of the ambiguity of these appointments expressed in shadow lengths. Let's say someone was invited to come to dinner when his shadow was ten feet long. Serenely disregarding the implicit reference to the setting sun, the moocher would show up early in the morning (pretending that he had understood that the morning sun was the reference) or even in the middle of the night (if he measured the length of his moon shadow).

The West Is the Land of Shadows

Tradition associates the solar bowl and its shadow calculations with a far more noble episode, one of the most important achievements of ancient science: the determination of the shape and size of the earth. The hero of this episode is the director of the library of Alexandria in Egypt, the astronomer, geographer, mathematician, and poet Eratosthenes (who lived from about 275 to 195 B.C.). Even though he did OK with all the sciences, he wasn't the best in any of them, and for this reason, apparently, his contemporaries nicknamed him Beta, or "Mr. Second-Rate." Eratosthenes knew that during the summer solstice the midday sun was reflected in the bottom of a well in Syene (today's Aswan, in Egypt). On the same day in Alexandria, 5,000 stadia (430 miles) farther north but at approximately the same meridian, objects cast a shadow, albeit a small one. This means that the sun is at its zenith in Syene and slightly lower near the horizon at Alexandria. How much lower? The graduated basin shows us that the little shadow sticking out from the base of the stylus covers about one-fiftieth of the circumference of which the bowl is a portion. Supposing that the sun rays are (practically) parallel, we can conclude that the 5,000 stadia of distance between Alexandria and Syene represents approximately one-fiftieth of the earth's circumference, which thus measures 250,000 stadia. By moving north or south, a sun clock loses its capacity to measure time, but it gains the capacity to measure space.

What we know from other sources about converting stadia into modern units gives us a distance that's very close to the real thing. This is remarkable, considering that the distance between the cities was measured according to the time it took a caravan to travel between them.

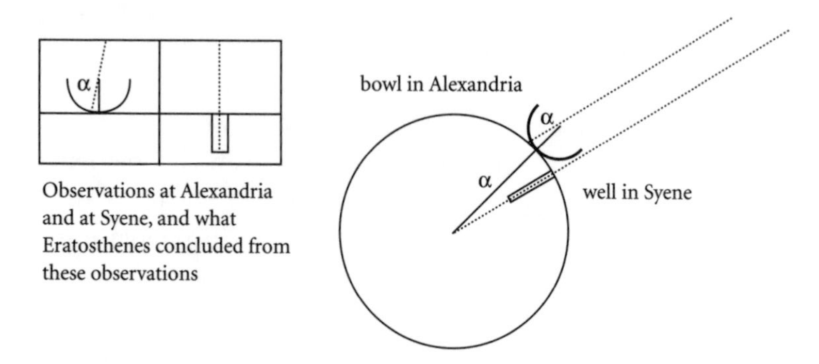

How to calculate the earth's radius using a bowl and just a little other information. Using the distance between Alexandria and Syene, the circumference turns out to be 250,000 stadia and the radius is 250/2π thousand stadia. The hypothesis is that the sun rays are parallel to one another.

This is one of the best-known experiments in the history of science, and not only for its simplicity and the accuracy and importance of the results. It looks like a magnificent example of how scientific research proceeds: first, observe; second, measure; third, state a hypothesis; and finally, calculate and determine your results. But there are some complications. For example, Eratosthenes did his reasoning on the basis of the hypothesis that the earth was spherical. If the earth were flat, the difference in shadows in Syene and Alexandria would depend on the fact that the sun is quite close in. (This is what happens to two people lit by a streetlamp, one of them nearby and the other farther away from the lamppost: they cast very different shadows because the light source is nearby.) And at this point the calculation would end up measuring the earth's distance from the sun.

There are other possibilities—the sun could be not so close, or the earth could be less curved—but we can disregard them. What matters is that we keep some basic hypotheses in mind. The distance between the earth and the sun has to be big enough that we can consider all sun rays to be parallel. Eratosthenes knew about this, in any case, from Aristarchus's research into the sun's distance from the earth. As we have seen, Aristarchus calculated that the sun's distance was somewhere between 380 and 1,500 terrestrial radii—really far away. Aristarchus's

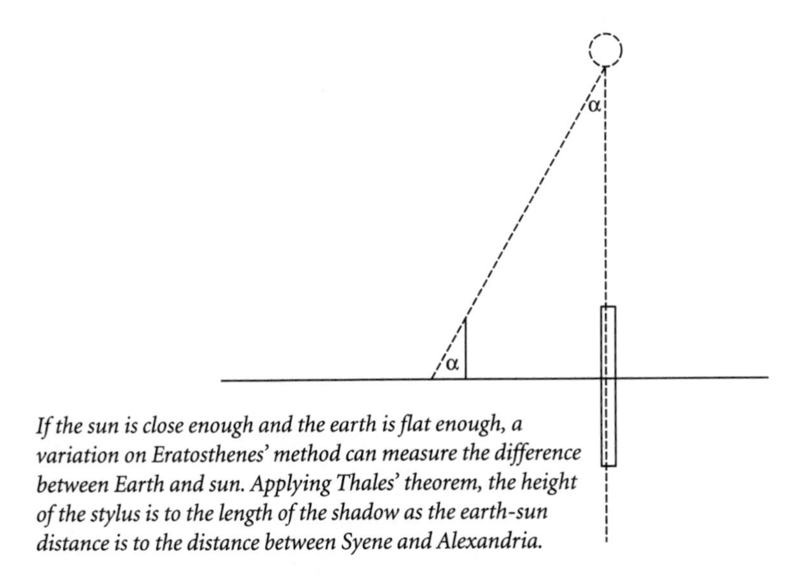

If the sun is close enough and the earth is flat enough, a variation on Eratosthenes' method can measure the difference between Earth and sun. Applying Thales' theorem, the height of the stylus is to the length of the shadow as the earth-sun distance is to the distance between Syene and Alexandria.

results were enough for Eratosthenes to presume that the sun's rays are sufficiently parallel to ensure that what he was measuring was in fact the circumference of the earth, and the accuracy of his measurement now depended only on the knowledge of the distance between Syene and Alexandria. And to get that measurement all you have to do is rely on the regularity of the speed of the caravans (or their camels).

Staying in the realm of the caravans and their affairs: once you know that one earthly degree corresponds to 700 stadia, you perform math of a more everyday kind. For example, you can determine how far you've traveled by checking the discrepancy between two shadows cast in the bowl (at the same time of day): you check the shadow first at your departure point and then again at your destination.

The Land of the Long Shadows

Eratosthenes' method lays a foundation for mathematical geography—as we can see from the strange poem below:

> Five zones were clustered around: two of them darker than cyan blue (lapis lazuli), only one arid and red as if from fire. This was central, it burned greatly as the flames struck it, so that it was inflamed by the rays which always warm things, since it stood right under the Dog-star. The things circling the poles on each end are

> always horrible (horribly cold), always wet with water, not water, but really ice that comes from the sky and covers the earth, the cold was extraordinary. . . . Two other [*zones of the earth*] are opposite one another in the midst of the heat and the rainy ice, both temperate, and they grow the fruit of the grain of Demeter Eleusina: in them men live at the antipodes.

As poetry, it's no great shakes: you can see why Eratosthenes was called Mr. Second-Rate. But it is an extraordinary synthesis of geography. What we have here is a description of the earth given by the god Hermes as he rises up to the sky: he clearly sees the climatic zones (the word "climate" comes from the Greek for "incline"). How did Eratosthenes know what the earth looked like from the sky? Well, as the director of the library of Alexandria, he had at his disposal a considerable portion of the scientific literature of the age, including various travel reports, which he synthesized in a book, *Geography,* that was consulted and copied for centuries after.

The most obvious way to organize geographic data is to diagram it in space. Eratosthenes had a fairly precise idea of the shape and size of the earth. Following the lead of Dicaearchus, a disciple of Aristarchus who lived a century earlier, he drew two basic lines that met at Rhodes: a parallel that passed through Gibraltar and the Himalayas, and a meridian through Syene and Alexandria. The other parallels and meridians would be found by studying the analogies among known climatic zones or the distances covered by travelers; but the most reliable method was found in the relationship between places on the earth and places in the sky. And here shadows are useful once again. Eratosthenes knew of the measurements made at Meroë by a certain Philo as he went up the Nile, as well as the reports by the greatest traveler of antiquity, Pytheas, who in the fourth century B.C. had measured the relationship between a stylus and its shadow in his hometown, Marseilles, and in other places he explored. Pytheas left Marseilles, perhaps to establish commercial relations with the people working the mines of Cornwall: he passed the Columns of Hercules, sailed up the French coast, circumnavigated Britain, and pushed north to a land where the sun never set during the summer solstice, the mythical Thule, whose precise geographic location puzzles commentators to this day. Many readers found the tale of his journey not to be credible, but geographers, including Eratosthenes, had faith in Pytheas, for they knew that shadows would be the key to latitudes.

The Land of the (Too-Short) Shadows

I'd like to draw attention to a shocking case of modern ignorance of Eratosthenes' method. Even without delving into the Thule myth, we will see that diffidence and disbelief are the order of the day at the extreme latitudes.

"There was no longer any east or west, there was no north; there was only one direction, south. Every breath of wind was a south wind. A year here consists of a single day and a single night. Had we spent the six months of the polar night here, we would have seen all the stars of the northern hemisphere describe their celestial orbits at the same constant distance from the horizon."

Beyond the mysterious Thule stretch the frozen wastes that surround the North Pole; more than two thousand years after Pytheas, at the end of the 1800s, the poles were the last parts of the planet to attract explorers. Financed by the Arctic Club of New York, Robert Peary tried unsuccessfully several times (in 1898, 1902, and 1906) to reach 90° latitude. On March 1, 1909, after having wintered at Cape Sheridan, he tried one last time with an expedition force of 24 men and 133 dogs. On April 1, while 240 kilometers from the pole, Peary selected his factotum Matthew Henson and the Inuit tribesmen Ootah, Egingwah, Seegloo, and Ooqueah, and pushed northward. On April 6 he planted the American flag on a pile of snow and started back south. Only on September 5 was he able to publicize the news of his conquest of the pole; he sent a telegram with the code word "Sun." Unfortunately, just five days earlier an American doctor, Frederick Cook, confirmed that he himself had reached the pole a whole year before, on April 21, 1908, and had been forced to spend the winter in the Arctic before he could get back to civilization. Public opinion was divided: the *New York Herald Tribune* supported Cook, while the *New York Times* and especially the National Geographic Society supported Peary. The controversy really heated up. The Parisian paper *Le Temps* feared that the debate over an entity as abstract as the pole was going to become an exercise in metaphysics. In the end, a political solution was found: on March 3, 1911, the American Congress voted in favor of the Peary Bill, declaring Peary the first man to reach the North Pole. But Peary never reached the pole. (It took the National Geographic Society eighty years to publish a retraction.) His reports contain many inconsistencies, and the photograph of the famous pile of snow betrays him. At the North Pole on April 6 the sun at midday is just barely 6° high in the sky. People and things ought to have very long shadows, almost ten times their own height. You might argue that Peary's men were modestly hiding their own puny shadows; but the landscape details are irrefutable and unforgiving. (Incidentally, Cook's photographs had the same problem.)

Robert Peary's team pretending to pose at the North Pole; they are betrayed by the shadows, which are too short.

Imprecise but Very Reliable: The Revenge of the Sundial

Let's come closer to the modern day and return to the subject of sundials. Apparently, Great Britain has more of them than any other country, despite its lack of sunshine—or maybe in stubborn defiance of its weather. (Does this demonstrate a romantic longing for the passage of time, or a desire to punish the elusive sun by putting it to work every time it shows its face?)

Whatever the reason, it can't be just a coincidence that the educational curriculum of the United Kingdom includes sundial study. Also for educational reasons, the planet Mars will have its own sundial, to be carried on one of the two Mars Exploration Rover missions, thanks to a project led by astronomer Woody Sullivan. Mars is fairly similar to

Earth, with a slightly longer day, and yet it's quite different, since its year is almost twice as long. The movement of Mars's shadows will be observed by a video camera linked to a Web site: there's nothing better for understanding how different Mars is from the earth than comparing our view with Mars's view of the sun's apparent movement.

These sound like great ideas. The humans starting the third millennium are a new brand, quite different from the humans who inhabited the earth before, even just a century ago. A hundred years back, Paris time was offset from London time by nine minutes and twenty seconds: each city marked noon at its own rate, and they vied for the monopoly on universal time. It was hard to find a solution; when found, it was then extended to the whole planet. The upshot—one of the first examples ever of globalization—is that nowadays we cannot simply depend on sundials, because local time differs from standardized time. The sun doesn't click in steps across the sky as it moves from one regulated time zone to another: every spot, every corner of the earth actually has its *own* midday. This didn't create major problems as long as the destination for most people's journeys was visible from the top of the local church steeple. But when in the 1800s traveling became both faster and more common, it was convenient to make time zones uniform and people forgot about sundials. More recently, the Swatch corporation launched the idea of a new universal time dividing the day into one-thousandths, in order to avoid references to time-zone changes and the annoyance of frequently resetting one's watch. It was a tentative try: because the numbering begins again with each new day, it is still based on somebody's local time—in this case, the local time of the Swatch company's own Swiss city. What would be really courageous would be a completely arbitrary counting system, starting with a date chosen at random, perhaps somewhere in the future to avoid favoring some birthday or other.

A hundred years ago not many people owned a clock, and finding out the time meant thinking for a moment about the sky's cycles. In the century since, we have agreed upon a mean time, and billions of clocks have been sold. But let's not forget that Greenwich Mean Time, however useful, is a fiction for everyone except those living right on the Greenwich meridian (it's a fiction even for their neighbors who live just a few inches west of the meridian). The only access to time for all the earth's other inhabitants is through our clocks synchronized to the sun's passage over Greenwich. Of course it's a good thing that there are billions of clocks around: if my watch stops, I can just ask a passerby for the time. But we have absolutely no idea what are the consequences of our complete dependence on these epistemic artifacts that measure such an important quantity in such an indirect way. A great technological blackout (however unlikely) would force us to reset our clocks by looking to the stars; and to do this we would have to consult a professional caste of astronomers. I bring up catastrophes just to make a rhetorical point: my aim is to emphasize that we have practically lost our direct knowledge of time based on the ability to read shadows and understand their movement—knowledge that was acquired with difficulty, was distilled through oral tradition, and was handed down through the centuries.

In some ways we are more guilty of ignorance than the Romans deplored by Pliny, because we have chosen ignorance. The sundial they stole from Catania hindered them from measuring the time in Rome, but today our wristwatches keep our eyes fixed on Greenwich and distract us from our own midday. The sundial, however, is poised for its revenge. Being a tool that marks astronomic cycles, it is particularly reliable because the astronomic cycles themselves are what make it tick. Even though it's less precise (because the shadow is always slightly fuzzy), a sundial is paradoxically more reliable than an atomic clock. The earth's rotation is not always regular. While an atomic clock proceeds impassively through this irregularity, it must—despite its technological superiority—still bow to the truth of the sundial. And sundials can be impertinent; as one inscription warns us:

> *Remember that we both count time as the shadow soars:*
> *You count out my passing hours, and I count out yours.*

IX In the Shadow of the Minaret

He causes the dawn to break;
and He has made the night for rest,
and the sun and the moon for reckoning.
—The Koran

Beyond scientific curiosities and practical utility, knowing about the relationship between the earth and the sky can help you out of some delicate situations. In the year 1024 in Ghazna, the most beautiful city of Afghanistan, the young sultan Mahmood, conqueror of eastern Iran and of the Punjab, received a delegation of Turks from the Volga who had come into contact with inhabitants of the polar regions. The ambassadors told the sultan that in the north at some times of the year the shadows grow very long and the sun doesn't set for several days. The sultan flew into a rage: this was heresy. The diplomats trembled. Then a wise man of the court intervened and patiently explained to the sultan all about the curvature of the earth and the whys and hows of the long polar days. The ambassadors' lives were saved. The wise man rejoiced—once again he had cleared away the mists of ignorance. But he knew that it was just one battle in an endless war.

The Shadow Man

The wise man was al-Biruni. The historian David King describes al-Biruni as the greatest of the Arab scientists. According to another historian, "The work of generations will be required to do full justice to Al-Biruni." The master *(al-Ustadh)*, as his contemporaries called him, wrote one of the greatest treatises on the astrolabe, a treatise on the calendar, various compendia of Arab astronomical knowledge, and descriptions of lands visited during his peregrinations—all of them works that unfortunately did not find their way to the Latinate West. Al-Biruni lived the life of a philosopher and a wandering astronomer. He was born in 973 in Khawarizm, an area south of the Aral Sea that changed hands several times, forcing him more than once to emigrate and start from scratch. Even as an adolescent he was mad about astronomical instruments; he constructed several of them for determining latitude. In 997 he observed the lunar eclipse of May 24 at Kath (his birthplace, to which he had returned after having fled it during a civil war). He had an agreement with an astronomer in Baghdad: from the discrepancy between the local times of the eclipse, the two of them could calculate the difference in longitude between Kath and Baghdad. The sultan Mahmood conquered Kath in 1017, and al-Biruni had to abandon the city once again. Tirelessly, despite the precarious conditions of his life, he managed to use an improvised sextant to measure the latitude of the village near Kabul where he took refuge. At the time, Mahmood's empire included part of northern India, where al-Biruni traveled far and wide. He learned Sanskrit and wrote a book on India. Al-Biruni hated sanctimony and ignorance; he despised the Arab conquerors of his native Khawarizm because they destroyed ancient books. This didn't stop him from learning Arabic and Persian, for he considered his mother tongue to be insufficient for expressing scientific concepts. Between 1030 and 1040 al-Biruni served Masood, and to him he dedicated the astronomy treatise known as the *Masoodi Canon.* He died around 1050; many years earlier he had dreamed that he'd live for another 170 moons, and in fact he seems to have done just that. He left 146 works, including *The Exhaustive Treatise on Shadows,* the most important book on shadows ever written.

Clearing Away the Mists of Ignorance

"Then there's a group of muezzins who come from those vulgar people whose hearts are disgusted by any talk of shadows, trigonomic functions, or altitude, and who get goose bumps if you so much as mention some calculation or a scientific tool. So much so that you can't even trust them about these topics, or even less about the proper hours for prayer; not because they are infidels or traitors, but because of their enormous ignorance."

The Exhaustive Treatise on Shadows serves primarily to solve religious problems. He who disregards astronomy risks distorting rituals, like the muezzin in the following story:

> For example, one of them came to me to ask for advice, and because of his enormous professional incompetence he obliged me to try and save him, for he was just guessing every time he used an instrument to determine doctrinal prayer times, and I was afraid that he would make a mistake in the rules of my religion. I explained Byzantine months to him. . . . At that point he began to suggest that Arab months should be used instead. So I explained that it has nothing to do with Arab months, and that aside from being completely confused, Arab months would require intercalation [the insertion of days or months to fix the gaps between the lunar and solar calendars], which is prohibited in Islam, which is considered absolutely heretical. But in the end his ignorance made him refuse anything based on the Byzantine months and he prohibited their use in the mosque, for fear of resembling the non-Muslim people. At that point I told him: the Byzantines also eat, and they walk in the marketplace. So don't imitate them in these two ways either!

Indeed, Islam, more than any other religion, seems to require constant attention to heavenly movements. For example, the five ritual prayers (at midday, in the afternoon, after sunset, at nightfall, and before dawn) demand that the moments of the day be clearly determined, and this determination must be respected to emphasize Islam's difference from pagan rites (some of which, for example, prescribe sunrise and sunset prayers that are probably residual traces of sun worship). Furthermore, the need to differentiate Islam from other religions (particularly Christianity and Judaism) resulted in the fact that the Islamic calendar is based strictly on the lunar month; but this makes the calendar continually slip out of phase with the solar year (in a solar year there are about twelve lunar months and eleven days). And finally, the start of the annual month of fasting is determined by the first moment that the waxing moon is visible; and even though it's not doctrinally legitimate to substitute astronomic reckoning for observation, calculations still facilitate the sighting of the crescent moon, because they allow observers to pinpoint the spot in the sky where it can be seen.

Al-Biruni tried more than once to convince his readers not to underestimate these problems, as did the muezzin in his story. To help them he tried, for example, to organize the imams' differing views on prayer times. There was quite a lot of discord about this, partly because the Koran is vague about its prescribed rituals, and partly because there are different traditions in Islam, and some religious authorities have said their own things, muddying the waters even further. One of the criteria for prayer times goes back to the tradition of the meeting between the Prophet and the archangel Gabriel in Mecca. The Prophet says:

> Actually Gabriel came with me twice to the door of the Kaaba and we recited . . . the afternoon prayer when the shadow of any thing was equal to that thing [when the sun is at 45° in the sky]. . . . The second day Gabriel recited the afternoon prayer with me when the shadow of every thing was twice as long as that thing [when the sun is at about 26° in the sky]. . . . The prayer times lie between these two extremes.

What happened? On the first day the angel showed the Prophet the earliest moment for starting prayers; on the second day he showed the time before which prayers must end. The afternoon prayer poses the most problems. According to this prescription, the prayer must be said between the time when the shadow is as long as the stylus casting it and the time when the shadow is twice as long as the stylus. But sometimes it's not possible to determine that earlier moment for every point on the earth's surface. As you go farther from the equator, the sun gets lower on the horizon, and in some places on some days the shadow is *always* more than twice as long as the stylus. (This wasn't clear to the sultan Mahmood when he grew irate with the Turkish ambassadors.) The practical rule followed by several Arab authorities in solving this problem was to start the afternoon prayer no earlier than the moment when the shadow is as long as the stylus plus the length of the shadow at midday. The rule can be applied everywhere except in Arctic regions. But it requires more than an elementary knowledge of geography and of the link between Earth and sky. For this reason al-Biruni strongly urged the muezzins to get to know the authorities—Archimedes, Apollonius, Euclid, and Ptolemy.

There's just one little technical problem: how can you determine the exact moment when a shadow is the same size as what's casting it?

Al-Biruni's geometric and systematic spirit wasn't hindered by any difficulty, no matter how surreal. For example, how about the head of the person casting a shadow? Because heads are round, the top edge of the shadow is cast by the point on the forehead tangential to the sun's rays. So it's not the highest point on that person's head. This complicates the measurement. Without meaning to, the cynical people of Khawarizm solved the problem once and for all: "They flatten the heads of their newborns and widen them in their cradles by pressing on the front and back, making them disgusting and turning them into warning objects for other peoples." The idea was that their disfigured children wouldn't be desirable slaves for their enemies. This ancient custom—Hippocrates too spoke of the Macrocephalics who lived around the Aral Sea—could turn out to be counterproductive: what if their enemies chose to use the natives of Khawarizm as sundial markers?

The Freezer of Shadows

Al-Biruni knew that a penumbra is just an unfocused shadow. Light sources are almost never a single point of light, so the farther away a shadow-casting object lies from its projection screen, the more indistinct is its shadow. In this context al-Biruni presented a curious phenomenon and evoked a strange theory of shadows that allegedly went back to Plato himself. This is the phenomenon: if you cast a shadow in profile and reach your hand toward your nose, the shadow hand will touch the shadow nose *before* your hand touches your nose. (Try it if you don't believe it. It's quite surprising.) "It is said that Plato said in the book *Timaeus*—when speaking about matter—that matter is shadow amid shadows, and that shadows emanate from objects and are frozen by a highly spiritual mechanism that condenses them into shadows." Al-Biruni's sarcastic comment is that "for these people the power of shadows is obvious. . . . According to what they say about Plato, a shadow . . . ought to be denser in the winter, and more rarefied in the summer, which is obviously stupid." Naturally, Plato never wrote anything of the sort, in the *Timaeus* or anywhere else; for the moment we'll take note only of the warning: never transform shadows into material things. Nevertheless, one mustn't rush to the other extreme and declare that shadows have no power at all, as did one al-Nashi, who actually dared to deny that eclipses were due to shadows: because shadows are incorporeal, he said, they can't hide the moon. "It was never more apt: anger and haste are the daughters of Satan," remarked al-Biruni.

☪

The recalcitrant muezzins of al-Biruni's time might have been won over by the series of anecdotes and shadow-lovers' curiosities. Leafing through the pages of the *Treatise*, we discover that the Indians made a rudimentary sundial by bending up a middle finger and measuring the length of its shadow along the palm with the fingers of the other hand. And that using a person as the stylus of a sundial was quite common (so much so that language itself testifies to it: according to E. S. Kennedy, the modern commentator on al-Biruni, "the oldest Sanskrit term for the shadow of a gnomon is *paurusi chaya*, or 'shadow-man'; this term was in use in India from the fourth century B.C."). We also learn that some counting systems are in base 7 because this is the number of feet necessary to measure the height of a person when his shadow is the same height as he is; or even 7½, to take into account the foot that sticks out only halfway from the vertical body.

☪

Al-Biruni also cited poetic passages regarding shadows: Midday, for one poet, is the moment in which "you walk on the nape of your neck." For another poet, it's the moment in which "the sun eats its own shadow, as the fire devours the twigs."

X *Time Flies Out Through the Hole in a Shadow*

Like Peary the explorer, the master engraver Giovanni Battista Piranesi (1720–1778) was betrayed by shadows. In his view of St. Peter's Basilica at the Vatican (overleaf), the light seems to come from the right side of the piazza. Well, the right side of the image is the north side of the square, which means that the sun is shining high in the northern sky of Rome. But Rome is in the northern hemisphere, where the sun *never* appears in the north. The obelisk at the center of the picture, brought to Rome from Heliopolis by Caligula in 37 B.C. and originally installed in Nero's Circus, was set in the piazza in 1586 under the direction of Egnazio Danti, a strange man of shadows. Danti wanted to use the obelisk for a sundial, and he would surely be horrified by Piranesi's astronomic and geographic blunder.

The obelisk that Danti wanted to make into a sundial in the Piazza San Pietro had an illustrious predecessor in the great solar clock constructed by Emperor Augustus (63 B.C.–A.D. 14) in Rome's Campus Martius. (The obelisk used there as a stylus is now in the Piazza Montecitorio.) This clock was a colossal work of propaganda reminding the Romans that Augustus had a special relationship with time. On his birthday—which in a magnificent coincidence fell on the day of the autumn equinox—the tip of the shadow ran in a perfect straight line toward the Ara Pacis, another Augustan monument. Augustus and Danti are linked across the ages because each of them used their works to celebrate a major reform in the way time was measured.

What's wrong with this picture? In the etching by G. B. Piranesi the light comes from the north *(the basilica is at the west of the piazza, so the shadows reach southward).*

The Tower of the Winds

I step out of the sunshine of the Piazza San Pietro. Entering the dimly lit room, I must pause for a moment so my eyes can adjust. Even though I know nothing bad can happen to me here—even though I know I'm among friends—I have the same impression one always has when stepping into a dark room. My body is tense and alert, and I feel as if I have antennae coming out of my eyes that are trying to push back the darkness bit by bit. The faint light in the room comes from a tiny hole about 5 meters up the south wall. From the hole a beam of light shoots to the ground, making a small, bright oval that moves past my feet slowly but inexorably. I begin to see markings on the floor, and in particular I see a *meridian* line. In just a few minutes the disk of sunlight will cross this line. Every now and then I can see the image of a cloud passing over the oval of the sun projected on the ground.

The oval is an image of the sun's disk. It's a hole in the room's shadow. Now I can make out the shape of the room: it's roughly square, and I see that the walls are frescoed; I glimpse some disturbing figures in the paintings. When the lights are turned on, the paintings show outdoor scenes of a stormy sea. The hole through which the sunlight shines is in the mouth of an old bearded man who pulls a cloud aside and seems to blow the light into the room.

I'm in the Torre dei Venti (Tower of the Winds) in the Vatican, built between 1578 and 1580 overlooking the Belvedere and the Pigna Courtyards—just outside the range of Piranesi's illustration. The room is probably the most valuable weathervane ever invented. Centered on the meridian line on the floor is a compass of wind directions that is replicated on the ceiling where a weathervane once spun to indicate the scantily clad figure corresponding to whatever wind was blowing. The walls are painted with fake draperies showing other meteorological allegories. The themes of the frescoes and the meridian project itself were the brainchild of Egnazio Danti (1536–1586), a Dominican friar and onetime mathematician and cosmographer to the grand duke of Tuscany, as well as a mathematician at the University of Bologna, who created this room shortly after entering papal service and becoming a member of the commission to reform the calendar. (He is also responsible for the designs for the Vatican's Map Gallery.) The wall illustrations are not merely scientific. Danti wrote: "On the south I had St. Peter's boat painted as it is being tossed by the waves; opposite I had the north wind painted, which represents the northern heresiarchs. The wind beats against the cliff and brings out the other winds in order to unleash them all against that most sacred boat which is, however, defended by the presence and the care of the Savior, who keeps it unharmed through every storm." And indeed, on the north wall is the inscription *Ab aquilone omne malum:* All ills come from the north. (The inscription was carefully removed during the 1655 visit of Queen Christina of Sweden, recently converted to Catholicism, who as a guest at the Vatican was housed in the Torre dei Venti.)

Blowing in through the hole in the southern wall is Austro, the south wind that clears away the clouds and brings peace back to the stormy seas that the Church was sailing during the Reformation. Not just wind but light comes in through this hole—the ray of light that disperses the shadows and keeps its daily appointment by hitting the meridian marker precisely each noon. But this appointment was regularly missed in the early years of the Torre dei Venti, because human time (measured by Julius Caesar's calendar) was out of sync with astronomical time (measured by the sun passing across the meridian) and the tower's meridian mark, disappointingly, bore witness to this problem.

The Legend of the Meridian

A millennium and a half earlier, when Caesar took the advice of various wise men, including Sosigenes of Alexandria, and introduced his Julian calendar, it was agreed that there were 365¼ days in a year. But the tropical year (the stretch between two successive passes of the sun through the spring equinox) was eleven minutes shorter. Mounting through the years, the error had built up into almost three whole days by the time of the Council of Nicaea (A.D. 325). The correction made during the council did not take into account the problem at the root of the mistake, and the calendar continued to go adrift for another twelve centuries. (In the fourteenth century the error made the winter solstice fall on December 13.) On February 24, 1582, shortly after the Torre dei Venti was built, the papal bull *Inter Gravissimas* corrected the time reckoning by skipping ten days and introducing the Gregorian calendar that we still use today. Legend has it that the meridian of the Torre dei Venti persuaded Gregory XIII that the calendar needed reorganization, but this is just a tale. The problem had been known for centuries, especially in ecclesiastical circles; there was lengthy debate about it at the Council of Trent (1545–1563), and in Danti's time any slightly educated person knew that conventional human time no longer agreed with sky time. Furthermore, a meridian like Danti's alone wasn't enough to convince anyone. The sun's passage during the year has only two notable spots, at the solstices, when the sun changes course in the sky; but at those points the sun's apparent movement slows down so much that the change in direction is imperceptible. It is most precise at the equinoxes, but unfortunately these are not particularly visible points: only a calculation can determine their position on the meridian line. So whoever was able, in those days, to appreciate the hidden mathematics determining the equinox point of the meridian also knew that the Julian calendar was out of sync. Danti's meridian cannot have added anything to the question of calendar reform (just as Thales' procedures cannot have contributed to the solution of the pyramid problem).

If you want to reform a calendar, it's a waste of time to try to convince an educated person by drawing a nice meridian line for demonstrations. The chief problem when trying to introduce such a major change isn't the accuracy of the reform but its *legibility.* You have to convince the largest possible number of people that they're wrong about a basic issue, that it would be better for them to change their minds, and that they have to do so in perfect coordination with lots of other people whose intentions they don't know and who might, perhaps, hinder the reform. (A similar problem is faced when a country where people drive on the left side of the road decides to change to right-side traffic. This reform cannot be made piecemeal: it works only if everybody agrees to start driving on the right all at once.) To get results you must find a mechanism that simplifies the cognitive effort for the people you need to persuade. The Gregorian reform was a success (apart from the irritating suppression of ten days, made to recoup the lag behind the sun) because it came from the pope, an internationally recognized authority; and also because it was so simple. The reform required a year (after the correction of the leap years) that produced an extra one-third of a day every 133 years, which by good luck meant one whole day every 400 years. (This leaves a residual discrepancy of one day every 3,327 years—a problem that we can leave for the next council to correct.) The calendar reform is a fine example of how an opinion gets spread: through pressure from the authorities, but also by the idea's intrinsic ability to attract support. This ability finally prevailed, even where the authorities had no effect or, indeed, were counterproductive. Antipapist Great Britain adopted the Gregorian calendar in 1752—with the suppression of eleven days, not ten, because the nation had accumulated a greater lag as it went on using the Julian calendar. Russia had to wait until 1917, and the "October Revolution" really began in November.

So the meridian of the Torre dei Venti is really only a decorative object. But we can certainly imagine the pleasure of the calendar reformers, once the change was made and they saw the sun positioned in the proper place at the proper time, marking the rediscovered accord between things human and celestial.

Some Good Reasons for Building Meridians

Danti is known for his other meridians. There is an extant one on the façade of the Church of Santa Maria Novella in Florence. His grandest construction, created in 1575, is found in the Basilica of San Petronio in Bologna, and it was so impressive that it made him famous and won him the position of papal cosmographer that brought him to Rome. This meridian no longer exists, because in 1653 the wall with the hole was demolished. In its place the astronomer Gian Domenico Cassini (1625–1712), at that time a professor of astronomy at Bologna, built another meridian. Cassini's *heliometer,* as he called it, was remarkably useful astronomically. His observations brought into question the reckoning of the obliquity of the ecliptic—that is, the inclination of the apparent orbit of the sun in the sky, during the year, with respect to the earth's axis. (We don't know when the first systemic comparisons were made between the length of shadows in summer and in winter, but we do know of an ancient shadow discovery: that the plane on which the sun seems to move is inclined with respect to the axis of apparent rotation of the stars.) Theoretically, this number can easily be found: you take the maximum value (summer solstice) and the minimum (winter solstice) of the sun's height in the sky, you measure the angle between them, and you divide by two. But the measurements are difficult: you need big meridians, as in Bologna, where the hole in the ceiling vault is 27 meters above the floor, and the sun runs more than 50 meters along the pavement between one solstice and the other. Cassini noted a significant discrepancy between the value he found and the one he expected, and he attributed this in part to the different refractions in the atmosphere.

Apparently, it boosts an astronomer's career to build a meridian in San Petronio. Thanks to his measurements, Cassini was called to Paris, where he became director of the observatory then being built. Changing his name to Jean-Dominique, he stayed there until the end of his life, and there he discovered Jupiter's red spot, four satellites of Saturn, and a gap between the rings of Saturn that now bears his name. It's only fair that Cassini, passing through Bologna in 1695, restored the meridian that made him famous.

That meridian is still an impressive tool. Cassini had to comply with a simple constraint—to trace the midday line from north to south. The Basilica of San Petronio is off the meridian's axis, so the line makes its own way, straddling two aisles. The meridian violently intervenes in the cathedral's rhythms: celestial geometry is indifferent to architects' creations. The light just barely manages to pass obliquely between two columns, like a slaloming skier; during the winter solstice it ends its journey by sliding a short way up the inside wall of the façade.

Like an undercover anthropologist, I mingled with the student groups and tourists one day to observe their reactions to the sun rays' movement at midday. They were stunned by the speed at which the sun's image moved; the inexorable appointment raised expectations that turned into amazement when the viewers realized that the sun does not actually stop on the meridian line. It's as if they expected the sun to halt and give us a moment to admire its alignment—but as soon as the thought has crossed your mind, the shining disk has already crossed the line. A man who seemed to know all about it didn't mince words as he grumbled about the imprecise explanations offered by well-meaning schoolteachers.

I hear that this man comes here every sunny day, as if he wanted confirmation of the stability of the cosmos.

Celestial regularity is always amazing. Giovanni Paltrinieri, a Bolognese builder of sundials, told me about a session of the city council called to approve his plan for a meridian in a public place. When Paltrinieri confidently reported to the council members that on the day of the solstice the sun would trace a particular line in a certain neighborhood piazza, more than one worried member asked him if it would really happen every year.

The same fascination with solar appointments draws numerous tourists who gather annually during the equinox to see the shadow of a serpent take shape at Chichén Itzá in the Yucatán. The stepped corner of the Pyramid of Kukulkan casts a serpentine shadow on the sloping side of the building; as the sun moves forward, the snake seems to slide across the wall, marking the passage of the season.

Shadows, tracing their patterns day by day around the objects that cast them, have lazily measured time for entire centuries. But in 1610 they were rudely awakened from their sluggish torpor and thrust into a flurry of belligerent declarations, of coded messages, suspicious glances, and great battles between scientists. The 1600s were the century of the shadow wars.

Part Three THE CENTURY OF SHADOW

(Curtain rises)

Plato and His Shadow

The light is strong but hazy. You can now sense that the sea is nearby. The sun is no longer unbearably hot. Skia, Plato's shadow, is growing by leaps and bounds, much more bold now.

PLATO: We've marched a long way through history, but we haven't made very much progress.

SKIA: Aristarchus and Eratosthenes show us how big the earth is; philosophers and mathematicians capture the sun's movement and make a clock from it; and you say we haven't made very much progress?! What do I have to do to persuade you?

PLATO: The shadow discoveries that you mention seem to have happened by chance. No one is giving you much credit.

SKIA: I've got even better things in store for you. There's one century—*my* century—in which all the astronomers will chase after me and woo me; they'll make a big deal of having caught sight of me here and there; they'll try to lure me out on the farthest planets of the solar system.

PLATO: Solar system? What's that?

SKIA: Times are changing, Plato. Thanks to me, people will know that the earth is not the center of the universe.

PLATO: Are you joking?

SKIA: Not at all. Read on. The best is yet to come.

XI *Shadow Wars*

OY!

A shout echoed across the sky: these were the two extra letters at the end of the anagram with which Galileo Galilei announced—in a letter of December 1610 to Giuliano de' Medici—the most important astronomic discovery since the time of the Greeks. The other words don't mean much by themselves: *Haec immatura a me iam frustra leguntur* (These are at present too young to be read by me) is an anagram. The solution was a hidden phrase describing a shadow tale that was to overturn the notion that had lasted from ancient astronomy: the idea that the earth is at the center of the universe, with the sun and the planets revolving around it. Now, in the seventeenth century, immense and spectacular shadows appeared on the scene, sparking many battles among scientists—and these shadows decided those battles by slicing through the universe like sabers. Shadows had the last word about the moon's real shape, about the nature and location of Venus, about Mercury's size, about the structure of Saturn's rings, and about the beat of Jupiter's satellites. But let's begin at the beginning.

Galileo's Moon

Galileo fired the first salvo in these wars. In 1610, just a few months before the letter with the anagram, he published an astronomical brief, the *Sidereus Nuncius (Starry Messenger),* with this drawing of the moon:

Galileo's pockmarked moon, from the Starry Messenger.

It's quite clear that the shadow line is irregular. But even more important are the white dots slightly to the left of the line at the center and the bottom. And to the right of the shadow line, some dark spots seem to be separating off from the shadow. Galileo had no doubt that the white dots were mountaintops emerging from the lunar night and being touched by the first rays of the sun, and that the black spots were the bottoms of craters still filled with night's darkness. In Galileo's words, "the moon is not covered with a polished surface but is rather rugged and uneven, and, just like the face of the earth, it is full of great outcroppings, deep cavities, and ravines." This drawing revolutionized astronomy and marked a leap after thousands of years of stargazing.

To understand the novelty of the image, we can compare Galileo's illustration with the paltry drawing of the moon by Leonardo da Vinci, whose skill at observation and pictorial rendering are usually beyond reproach. The recto of folio 310 of the *Codex Atlanticus* shows an example of what Leonardo had to offer from his investigation of the moon. The play of light and dark does not record mountaintops lit by sunlight and valleys full of shadow. Leonardo saw nothing more than countless other observers had seen before him: a strange, sad, indecipherable face.

Galileo was fully aware of the novelty and the importance of what he was doing when he presented his drawings of the moon: "In truth, I propose great things in this brief treatise for the observation and contemplation of anyone who studies nature. I say great because of the excellence of the material itself and because of the novelty never offered before in all the centuries, and lastly because of the instrument by which these very things were revealed to our senses." The greatest novelty was naturally the new observation tool, the telescope. Galileo says that he had heard talk of it some ten months earlier. Based on a simple description of a telescope, he says, he was able to figure out the laws that

Leonardo's disheartened moon from the Codex Atlanticus.

allow such magnification. And then—with remarkable improvements—he replicated the instrument, "drawn from the most recondite speculation about perspective," as he wrote to the doge of Venice, Leonardo Donato, on August 24, 1609, when offering him a telescope. This claim must be taken with a grain of salt; it's likely that Galileo examined a telescope brought to Venice by a traveler from Flanders in 1609, and that he took it apart in order to perfect it. When he gave his telescope to the doge, Galileo immediately showed him its military utility: "On the sea one can spy the timbers and sails of the enemy from much farther away than usual." Its scientific importance was also immediately clear, its ability to expose the essence of celestial things "to the senses as well as the intellect." The telescope helps us to discover facts about the sky "with the certainty of the experience of the senses," and it thus allows us to settle some old questions that had migrated from the realm of astronomy to the realm of metaphysics.

The *Starry Messenger* was printed in haste in March 1610 with the bonus gift, in the dedication, of the four satellites of Jupiter—also discovered with the telescope—offered to Cosimo II de' Medici. (Cosimo was Galileo's potential protector; sure, it was a self-interested present—but who else could ever give stars as gifts?) The treatise quickly spread through the community of the educated and the royal courts of the whole known world.

The Visible and the Invisible

Naturally, Galileo's drawing will astonish many readers. The features of lunar topography are so rough; how is it that no one glimpsed them before Galileo? Doesn't the engraving represent what we would see by simply scrutinizing the moon with our naked eyes? Well, this objection is raised only because once again we are trapped by the difficulty of remembering the apparent size of heavenly bodies. When we discussed the earliest astronomical discoveries, we saw how the moon is generally exaggerated in the mind's eye. To see how the 1½-inch drawing at the start of this chapter corresponds to the moon that you see in the sky, you have to step back 14 feet from the book—and at that distance you won't see much in the picture.

But when Galileo pointed his telescope at the moon, he saw it "as if it were only as far as twice the earth's radius." It's as if you were holding this book only 6 inches from your nose and looking at the picture at the beginning of the chapter. So Galileo noticed immediately that the moon's surface is ruffled not only by the great spots visible at all times from anywhere, but also by many smaller impurities that are visible at the edge where light meets shadow; they arise, they change shape, and they disappear with the waxing and waning of the moon. These spots, invisible to the naked eye, appeared to him as a play of shadows and light, and they posed a problem of inverse optics for Galileo: to reconstruct the shape of the object based on the shape of its shadows.

Galileo was not only an astronomer. He was also a master draftsman, and he knew everything about shadows and how an object's shape is revealed by the changes in its shadows. And so that's how he thought about the moon. The shadow running across a perfectly spherical solid body must follow a smooth line, but the shadow line on the moon is "uneven, sharp, and quite sinuous. Indeed many luminous points, like excrescences, reach beyond the limit of the light or the shadow, and by contrast some dark parts intrude into the illuminated part." Then there are big blackish spots that invade the area lighted by the sun—spots that are dark on the side the light is coming from but that on the other side are "crowned with brilliant edges, almost like burning mountains." And there's more. The shadow line—and the distribution of the spots—are not the only things that change with time (because the moon spins on its axis). The division of light and dark evolves, but there is more to it: this evolution follows an organized plan. An analogy enters:

"We have a similar spectacle on the earth, toward sunrise, when we see the valleys not yet lighted up, and the mountains around them shining on the side facing the sun; and while the shadows of the terrestrial cavities grow smaller bit by bit as the sun rises, so too these lunar spots, with the growth of the luminous part, lose their shadows."

Galileo was the first to see the magnificent spectacle of the slow rising of the sun on the mountains of the moon, and he described it eloquently:

> Indeed, not only are the boundaries of light and shadow in the moon seen to be uneven and wavy but—still more astonishingly—many bright peaks appear within the dark part of the moon, completely divided and separated from the illuminated part and some distance away from it. After a certain time these increase in size and brightness, bit by bit; after two or three hours they join up with the rest of the lighted part which has now become larger. In the meantime more and more peaks shoot up as if sprouting—now here, now there—lighting up within the shadowed portion; they become larger, and finally they too join that luminous surface which extends ever further.

Galileo's reasoning couldn't be simpler. Since they *look* like mountains, we have every reason to believe they *are* mountains. And monstrously huge mountains and valleys they are, much larger than those on the earth. A simple geometric demonstration led Galileo to conclude that lunar elevations might reach as high as 5 miles. (Galileo believed that there were no mountains on Earth higher than 6,500 feet, which led him to think that the lunar mountains were higher than those on Earth.)

Shadow Games and Crystal Moons

Galileo's huge lunar mountains were hard to swallow. In 1611 the Jesuit Cristoforo Clavio (1537–1612) and a certain Ludovico delle Colombe asserted that shadows were only one possible interpretation for the spots visible through the telescope; they could also be seen as differences in the density of the material the moon was made of—they look like shadows of mountains, but it's not certain that's what they are. Delle Colombe looked through the telescope and admitted that the moon

Even from up close, shadows are useful. A ground-skimming shadow amplifies details and shows the shape of lunar mountains, as seen in this photograph shot by Apollo 12.

appeared to be a mountainous planet, "seeing in the lunar body many differences of thinness, which resemble mounts and valleys and crags, so that it seemed I was gazing at a second Earth." But he would not be persuaded. After all, even paintings suggest relief with chiaroscuro effects, and yet paintings are perfectly flat. "Where a body that is square appears to be round; the spherical, flat; and the flat, because of variety of color, shadows, and light, appears to have depths and relief: and nonetheless this appearance is false, as paintings show us." Delle Colombe's hypothesis was that there were different densities to different parts of the moon, "which densities are not only on the surface of that body, like colors in a painting, but also scattered inside that whole body, and they have all corporeal dimensions, being wide, long, and deep, in the same way that the mountains and valleys would be, if there were any in that body"; but the parts that cover these mountains and valleys are invisible because they are so thin. Delle Colombe used an analogy to explain: "A manifest example would be if someone took a large ball of very clear crystal inside of which was a tiny Earth of white enamel, with forests, valleys, and mountains; exposed to the Sun toward the Sky quite far from the eyes of whoever looks into it, this sphere doesn't look smooth and spherical but uneven and mountainous, and shadowed wherever the Sun is not.... But what is the point of seeking examples among inferior things, when it's quite clear up in the Sky itself?"

This interesting hypothesis has a problem. If the moon were lighter and darker not because of valleys and peaks, but because of the way that light passes through it, why in the world would the material that the moon is made of display *exactly* the trousseau of shadows that indicate valleys and peaks? What's the reason for the moon, with its smooth surface, to *imitate* a mountainous planet? Galileo himself ridiculed that objection, observing that the moon would anyway be imperfect if it had those strange mountains submerged in a sea of crystal. Another objector, a certain Sizzi, retorted that they really *didn't seem* like mountains, or at least that Galileo didn't do enough to convince people. And indeed Galileo repeated over and over that what he saw was exactly the configuration that corresponded to mountain shadows, but he never gave a geometric demonstration.

What determines whether a certain shadow configuration does or does not look like a mountain? Three centuries after Galileo, the astronomer Camille Flammarion (1842–1925) published *Popular Astronomy,* in which, in order to convince people that certain zones of the moon were volcanic, he played with shadows, placing a picture of a lunar region next to a picture of Vesuvius and the volcanic Phlegrean Fields around Naples. Ironically, the analogy between Earth and moon works in the opposite direction here: the picture of the moon is what you can actually see

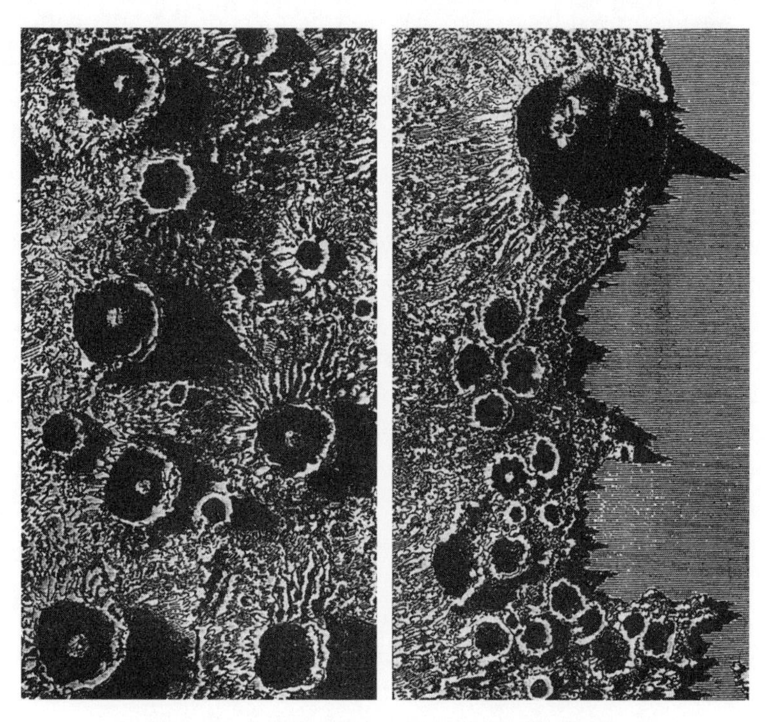

The moon (at left, seen through a telescope) is like the earth because it appears as the earth would appear *if seen from the moon (at right, a picture of how the Phlegrean Fields and Vesuvius would appear). But how did Flammarion get this picture?*

through a telescope, while the terrestrial picture is Flammarion's reconstruction (note Vesuvius's sinister shadow) based on how the earth would appear from the moon.

We will soon see how Galileo too tried to use a picture to convince people without using many words. For the moment we'll consider the consequences of his discovery. Why was it so important to know in 1610 that the moon was pockmarked? Why did Sizzi, Clavio, and delle Colombe go to so much trouble to try to demonstrate that it wasn't?

The Importance of Being Pockmarked

In the context of the astronomy of the time, Galileo's observations were important because they demonstrated that the moon was like the earth at least in shape. "Just like the face of the earth," wrote Galileo. To judge from what can be seen by telescope, we cannot say that the moon is smoothly polished. But that was exactly the astronomical orthodoxy that was disproved by Galileo. The astronomers of the time naturally had many reasons for refusing the new Copernican, heliocentric model (the sun at the center of the universe, with the earth spinning on its own axis) and sticking with the more intuitive Ptolomaic, geocentric model (the earth immobile in the center, sun and planets circling around it). One reason was the belief that heavenly bodies were radically different from the earth. The moon's status was uncertain: Was it too close to the earth not to be partly corrupted by it? Or was it pure and uncontaminated like the other planets? The presence of mountains and craters showed that the moon had no prerogative to perfection as a heavenly body, that it was not substantially different from the earth. The imperfection of the moon showed that there are not *two* worlds—the terrestrial and the celestial—with completely different laws. If the moon can be like the earth, the earth can be like the moon, and it can in turn circle around another celestial body. The earth is subject to the laws of the sky; it too is a wanderer.

The fact remains that it wasn't easy to understand that there were mountains and valleys up there. Even using the telescope, Galileo had to *interpret* what he saw. It's not true that one immediately perceives that the moon is mountainous: one must patiently collect a series of observations. Only then does one see that the spots, "like steep cliffs, and with sharply angled rocks, stood out from one another with clear contrasts of light and shadow," and that they appeared "because of the unevenness of their aspects as a consequence of the different illumination by the sun, which moved the shadows in different ways. . . . From day to day their aspect changes, they grow and shrink and disappear, because they originate only from the shadows of the upraised parts."

The telescope showed the difference between the modern spots and the old-fashioned scars that hint at a face of the moon: those scars had been visible to everyone and were scrutinized suspiciously by the ancients. The lighting of those older marks changed only slowly, and independent of the sun's rays. The new marks were not only invisible to the naked eye; they were also a different sort of thing.

Old and New Spots: The Face of the Moon

The prince of astronomy, Johannes Kepler (1571–1630), quickly responded to the publication of the *Starry Messenger* in an open letter dated April 19, 1610. He hadn't yet tested Galileo's discoveries, probably because he didn't have a strong enough telescope. He congratulated Galileo on his discoveries, but he took a prudent position, approving only the theories that seemed likely or that corresponded to facts that could be verified even without a telescope. For example, he considered Galileo's observations on lunar imperfections not as an original discovery but as the confirmation of what others had already said. These others were Democritus, Plutarch, Kepler himself, and above all Maestlin, Kepler's master in Tübingen, who in 1605 had asserted in a little book that "the moon and the earth are similar in density, shadow, opacity, and illumination by the sun that circles around both globes, so that the moon shows its phases to terrestrials just as the earth shows phases to the inhabitants of the moon; thus both bodies light each other reciprocally." In practice, Galileo hadn't discovered anything new.

But which point of Galileo's had already been made? The question is not only the moon's mountainous (hence terrestrial) nature, but especially the kind of proof that can be offered for this. The recurring proof in the literature before Galileo was that the lunar spots *visible to the naked eye* indicated dips or hollows. And on this point there is indeed a long history. For example, the characters in *The Face of the Moon* by Plutarch (A.D. 45–125), one of the classics of the ancient world (which, incidentally, was translated into Latin by Kepler), have a sophisticated discussion about the great spots: Are they shadows or are they not? If they are, then the moon is rugged and is similar to the earth; if not, then they are due to the thicker or thinner density of the material of which the moon is made. Lampria, the dialogue's narrator as well as the brother of and spokesman for Plutarch, defends the former hypothesis and eliminates each objection to the theory of a rugged moon. "It follows that we shouldn't think we're offending the moon by considering it earthly; and as for the face that appears on it, it has great depressions like our Earth so that the ground is wrinkled with subsidences and equally large fractures, actual reservoirs of water and dark air. Therein the sunlight which does not reach inside them and does not even brush them, fails, and reflects intermittently back to Earth." When his friend Silla joins the dialogue to tell a reassuring story about the moon being the home for the souls of the dead, the lunar chasms are revealed to have a function: they are the resting place for the souls of people who were evil in life. Plutarch's lunar purgatory foreshadows Dante's *Paradiso,* which again raises the question of the nature of the spots. At first Dante too shared the idea that the "blue marks" on the moon were due to a difference in "thin and dense bodies." But how to know what's really what? There's nothing better than asking someone who contemplates the stars from close up—Dante's guide, the soul of Beatrice. Beatrice answers him with a long explanation. (Sensibly, she notes in passing that if the moon were transparent there would be no solar eclipses.) Beatrice's theory mixes physics and myth, saying that credit goes to the intelligent mover of the universe: "From this proceeds whate'er from light to light / Appeareth different, not from dense and rare: / This is the formal principle that produces, / According to its goodness, dark and bright." This doesn't make it any clearer how things work, and I have the impression that Dante's commentators cannot agree on a meaning.

Except for Dante, the pre-Galilean protagonists of this discussion are all partly right about the spots, but it's hard to evaluate their arguments because no one had the right method for corroborating his position. Lampria is right—the moon is full of hollows—but to demonstrate it he refers to the wrong spots, and his explanation artfully throws the reader off track. Indeed, his interlocutor, Apollonius (who may be playing the mathematician of the same name who studied conic sections—another member of the family of umbraphiles), raises the objection that the shadows on the moon are so large that the corresponding mountains and depressions must be immense, and therefore visible from Earth. Lampria ridicules this objection, asserting that even a small object can cast an immense shadow. But his explanation takes into account only the lighted object's *distance* from the source, not—as it should—the opening of the shadow cones: "The shadow is magnified not by the size of the irregularities on the lunar surface but by the intrinsic distance of the light." A little bit like the chasms that can be seen, better than the mountains, from very far away when the light is diffused by distance. We could say that Lampria sees more shadows than there really are, and he has to stretch in order to explain them, comparing the darkness of lunar hollows to a phenomenon of aerial perspective. But this is nothing more than prestidigitation; and, given the solid competence he showed about the phenomena of shadow, it's clear that Lampria/Plutarch knows that he is lying when he says this.

Kepler too guessed the truth (almost), although for the wrong reasons. He forgot about the shadows, but he clung to the hollows, associating them with huge water basins, and he asserted peremptorily: "I declare that the dark spots are seas and the bright parts are land." Even today the great dark spots are known as seas. But wrongly so: we know that the so-called lunar seas are dark not because they're full of water but because they are large expanses of basalt, which has low reflectivity and thus appears black. (In March 1998 the data gathered by the *Lunar Prospector* probe led to a hypothesis that there was water on the moon. Where? In the only zone permanently in shadow—the bottom of the craters at the poles—in the form of ice. Because the moon has no atmosphere, the ice is believed to sublimate and evaporate as soon as the sun hits it, and because of the low gravity the vapor must escape into space. Every part of the moon is lit by the sun sooner or later, except for the patches of dirt at the bottom of the polar craters—all in all, these patches cover something between 2,000 and 6,000 square miles. The presence of water has not been confirmed; but if there are seas on the moon, they lie in Galileo's craters.)

To say that the moon has mountains, people have to look at the right spots, and that is possible only with a telescope. Galileo was right to talk about the moon's spikiness, mountains, and craters, because what he saw were true shadows.

What Galileo Really Saw: The Mystery of the Bohemian Crater

We know that Galileo wasn't the first to use a telescope to scrutinize the sky. The English mathematician Thomas Harriott (1560–1621) observed the moon on August 5, 1609, and made a drawing of it. But the drawing is incomprehensible. The shadow line in Harriott's drawing is physically impossible. All shadow lines in every phase of the moon must intersect the satellite's profile at two diametrically opposed points. Harriott's line pushes too far, like a child's drawing.

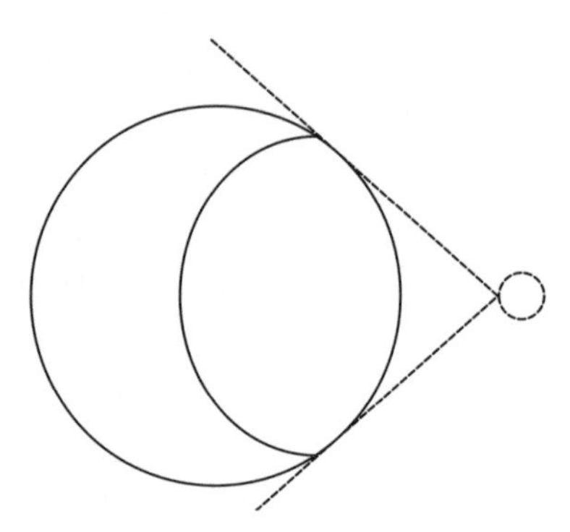

A sketch showing Harriott's impossible shadow line. I have added the sun to show its position if it were to create such a shadow.

Harriott's drawing is even worse than Leonardo's; at least Leonardo made it clear that it represented the moon. Galileo's drawings are not only more accurate; they also reflect the fact that Galileo understood what he managed to see. Galileo's wisdom lies in having grasped that the brightnesses and darknesses were a play of light and shadow. Indirect testimony confirms that shadows were the crucial element. William Lower, a collaborator of Harriott's, heard of Galileo's discoveries and wrote, "I had formerlie observed a strange spottednesse al over, but had no conceite that anie part thereof mighte be shadowes."

Despite Galileo's superiority, the drawings published in the *Starry Messenger* were not very precise. Let's look at the following illustrations:

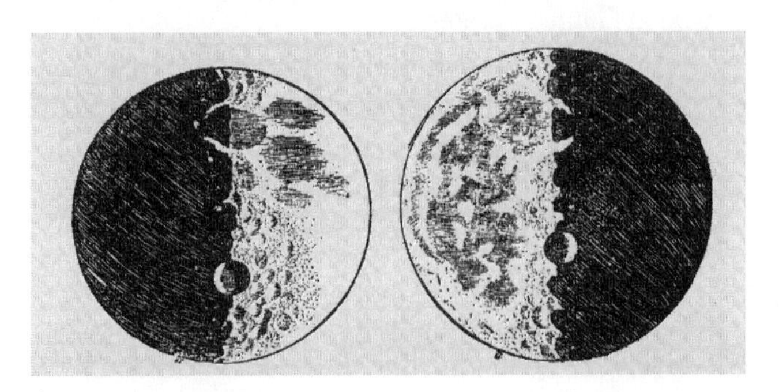

The first and last quarters of the moon in an engraving from the Starry Messenger. *The central crater doesn't really exist.*

Along the shadow line just below the center can be seen an enormous crater with a well-defined outline. Galileo spoke about this at length. "And I also want to record another thing that I noted with some surprise: almost in the middle of the moon is a cavity greater than all the others and perfectly round in shape: I glimpsed this in the vicinity of both first and last quarters." (That is, the shadow line passed through the same points in the two pictures, but the sun illuminates the moon in the picture once from the right at the first quarter and once from the left at the last quarter.) "And as best I could, I reproduced it in the two figures placed here above: in shading and in illumination it offers the same aspect that, on the earth, would be offered by a region like Bohemia, if it were surrounded on all sides by very high mountains disposed in a perfect circle." This lunar Bohemia has puzzled modern readers. It doesn't correspond to any characteristic of the moon known to be in that area. (It may be possible to identify it as the Albategnius crater, or as a series of craters grouped slightly below the Ptolemaeus crater, but the printed drawing exaggerates its morphology enormously.) Historians and philosophers from Arthur Koestler to Paul Feyerabend have criticized Galileo for his inexact map. But other historians, in Galileo's defense, have observed that Galileo's original hand drawings (below) are more faithful to the appearance of the moon than are the engravings of the *Starry Messenger*. In particular, these originals do not show the large Bohemian crater; or at least they don't show it surrounded by very high mountain chains. Thus, they say, what appears in print is an error.

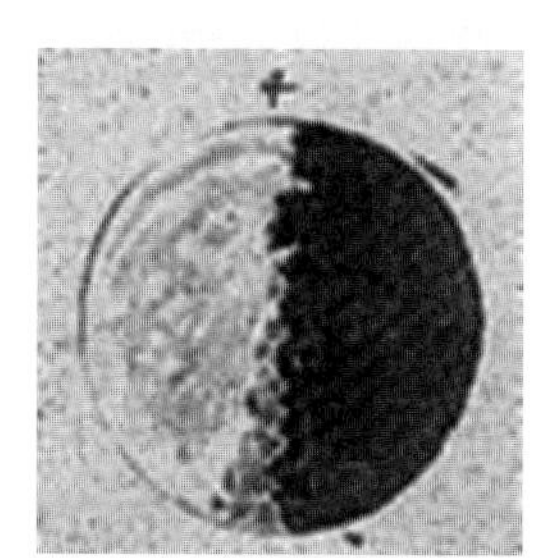

Galileo's original drawing shows no trace of the great crater.

Well, did or didn't Galileo properly represent what he saw through the telescope? Can the drawings published in the *Starry Messenger* be seen as proof of scientific fraud? And why would he perpetrate such a fraud?

To solve this little mystery we have to take some circumstances into account. To start with, one fact that seems to have escaped the historians' notice ought to be a central element in the debate. A crater with the depth and dimensions of the Bohemian crater *would be visible to the naked eye.* If you look at the engraving on page 124 while standing 15 feet away (to simulate the apparent size of the moon), you'll see that the Bohemian crater is quite tiny—but visible.

Galileo couldn't really hope to escape criticism from people who objected that they didn't see the crater on the moon's surface with their naked eyes. (Kepler declared it a marvelous discovery, but maybe he was being sarcastic.) And this suggests that in representing it so exaggeratedly in the printed picture, Galileo was concerned with something other than giving a good description of it.

The second circumstance that we should keep in mind is that the Bohemian pseudo-crater is shown in two of the four illustrations printed in the *Starry Messenger.* It's hard to imagine that the engraver made the same mistake twice while translating Galileo's sketches into illustrations. It really cannot be a graphic typo. The crater is noticeably lighted from the right in the first drawing and from the left in the second (following the progression of the lunar phases), almost as if to underscore the concept. Furthermore, Galileo himself described the lunar Bohemia as "surrounded on all sides by very high mountains disposed in a perfect circle." This description matches the printed illustrations and not the autograph drawings. Finally, the description of the crater appears immediately before Galileo's explicit recapitulation, according to which the configuration of the spots on the moon must be interpreted as a play of shadows and light, and as such constitute "incontestable proof of the sharpnesses and unevennesses" of that satellite. This provides the motive for the heavily charged—you might even say heavily caricatured—representation of the crater.

If we keep all these circumstances in mind, we come to the following interpretation. What appears redundantly in the printed edition is a *didactic* crater that offers an effective example of the shadow method that helped Galileo to understand the structure of the moon's surface. In the engraving the Bohemian crater is the only object whose shape can be understood at a glance, without additional explanation. It's as if Galileo were saying to us: this, as you can easily imagine, is what appears of any crater seen from above and lighted first from the right and then from the left, and the changing view of the moon's shadow line must be interpreted according to the model that you see here. The very picture of the crater makes us think of a moon within the moon, with shadows reversed as in Heraclitus's moon bowl.

So is the case closed? Not necessarily. Let's imagine Galileo working with his engraver. He has his watercolors in front of him and he is telling the craftsman how to proceed. As one always does in an artist's studio, Galileo holds the paper at a distance and squints to see if the shadings really give the impression of relief. Since the watercolors aren't really photographs of the moon, there is still some room for interpretation when translating the pictures into definitive illustrations. Look at what happens if we reproduce this process using digital blurring: the Bohemian crater appears as if by magic.

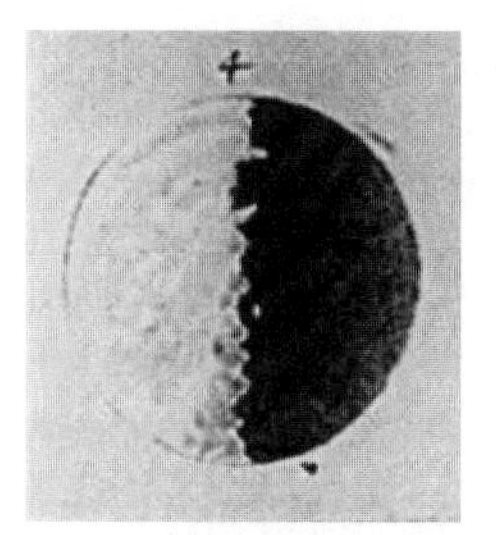

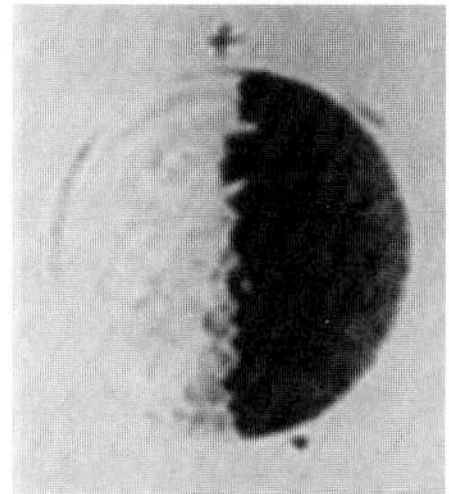

An explanation of the Bohemian crater error as a problem of perception. If Galileo's watercolor is "softened" (as when looked at through squinting eyes), a formation like a crater can be seen.

XII Venus Imitates the Shadow of Diana

I used to measure the heavens,
now I shall measure the shadows of the earth.
Although my soul was from heaven,
the shadow of my body lies here.
—Kepler's epitaph

The great astronomical discoveries based on shadow had to await the invention of the telescope. And the telescope made Galileo famous. Thanks to the *Starry Messenger,* Galileo obtained the coveted position of chief mathematician and philosopher to Grand Duke Cosimo II de' Medici, and he moved to Florence on September 12, 1610. Free of teaching obligations, and probably anxious to maintain both his place as top astronomer and the goodwill of his patron, he hastened to announce new discoveries.

The first of these, made as early as July, was about Saturn—but we'll get to that in the next chapter. In December 1610 he sent Kepler the anagram that we already know about, "*Haec immatura a me iam frustra leguntur o y,*" via Giuliano de' Medici. Disguising the news of one's discovery was a way to avoid divulging it before one was sure, but without relinquishing the option of claiming the invention if others tried to take the credit. (Galileo was prone to controversy.) Kepler tried in vain to decipher this message and finally wrote to Galileo begging him to reveal the solution: "I beg you to keep in mind that you're dealing with honest Germans." He was offered the key on January 1, 1611, in another missive to Giuliano de' Medici in Prague: *Cynthiae figuras aemulatur mater amorum* (The mother of love emulates the figures of Cynthia). The Greek "Kynthios" or "Kynthias," Latinized as "Cynthios" or "Cynthia," comes from the sacred birthplace of the twins Artemis and Apollo, the little Mount Cynthos of the island of Delos in the Aegean. It may be that Artemis/Cynthia is the goddess of the moon because she was the goddess of hunting: the moon's crescent is like a bow pulled taut in the sky.

So the solution to the anagram is: "Venus is like the moon, she imitates its aspects."

There are two parts to this discovery. Venus is like the moon because it doesn't shine with its own light—so its illumination comes from the light of another star, the sun. In Galileo's words, Venus is *tenebrous.* But above all, Venus—like the moon—has a *complete* cycle of phases. There's a new, a waxing, a full, and a waning Venus, and this proves that the planet *describes a complete orbit around the sun.*

It seems like a small thing, but it was truly the end of an era. In the Ptolemaic system, as in Aristotelian cosmology, Venus orbited around the earth and not around the sun, so these appearances of Venus could not be kept up. In fact, it was believed that Venus was *always* between the sun and the earth; even though a Venus that doesn't shine with its own light must have phases, it wouldn't be possible to see a "full Venus" (to look full, Venus would have to be *beyond* the sun).

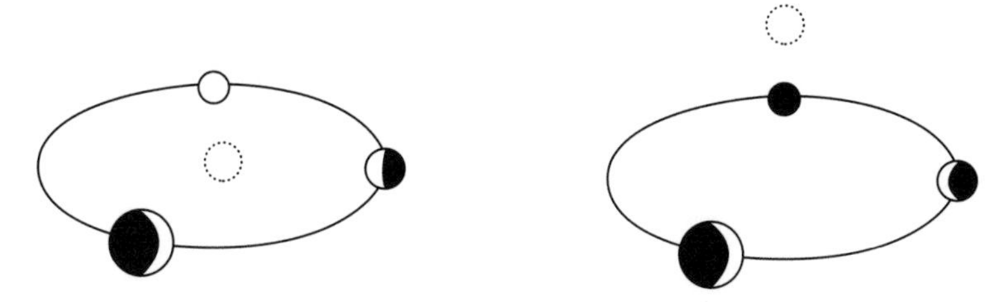

If Venus (seen from the earth; not to scale) is full, this means that it lies beyond the sun at some point in its orbit (drawing at left). If it orbits on this side of the sun (on an "epicycle" whose center orbits in turn around the earth), it can never be full (right).

The Ptolemaic system is wrong about this point: it can't explain Venus's aspects, even though the use of the telescope changed forever the notion of an aspect. Of course, it was quickly noted (and Kepler emphasized many times) that observation of the complete cycle of Venus's phases wasn't enough to confirm the Copernican system; for example, even in the geoheliocentric system of Tycho Brahe (1546–1601)—in which Venus orbits around the sun, and the sun in turn orbits around

Earth—Venus can seem to have the whole set of phases. But Galileo's observation certainly makes mincemeat of the Ptolemaic system. There were fewer things circling around the earth than were foreseen by the old canon of astronomy, so much so that after a short time the canon stopped being used by any professional astronomer.

Tenebrous Venus

The discovery that Venus didn't produce its own light was greeted with shock. Kepler was dumbstruck, and he wrote to Galileo to ask him to continue observing Venus to verify a certain hypothesis he had—that the planet was made of the finest gold. In the Aristotelian-Ptolemaic system of the universe handed down from medieval science, it wasn't clear whether the planets and the stars shone with their own light or whether they reflected the light of the sun. Aristotle envisioned a clear split between the sublunar world, shadowy and cindery, and the celestial shining bodies (with the moon as the only exception, because it was too close to the sublunar area not to be somehow corrupted by it). But the scholastic philosophers preferred as a rule to maintain that the light of the stars and the planets was parasitic to the sun. Observation with the naked eye, however, seemed to disprove this. If Venus is a luminous parasite of the sun, then why can't we see its phases, since it's always in a different position in respect to the sun and the earth? If it has phases, Venus should look brighter or dimmer, even to the naked eye, according to its position. This is the gist of the theory of Avicenna (980–1037), who thought that Venus had light of its own. In the second half of the 1300s he was answered (wrongly) by Albert of Saxony: "Venus and Mercury are so transparent that they incorporate the sun's light and are permeated by it, as different from what happens with the moon." Thus, Venus doesn't shine with its own light, but this doesn't mean that it's dark: like a frosted-glass sphere, it diffuses the rays of the sun uniformly, without casting shadows, and no part of it lacks light.

In hindsight, we know that the controversy could not be solved before Galileo. When full, Venus is so far from Earth that it turns us earthlings a lighted surface smaller than that shown by the waxing or waning Venus, and the difference is not noticeable to the naked eye.

When Galileo observed the phases of Venus, he reconstructed the relationship between Venus and the sun just as Aristarchus and the early astronomers had with the moon, and he closed out the first great period

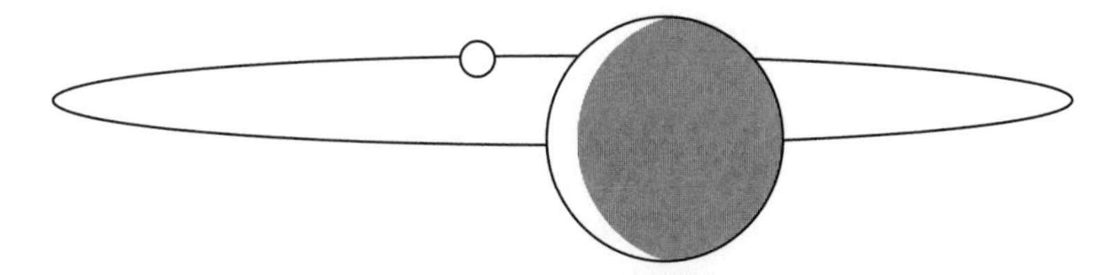

Venus when it's full (in the background) is much farther from Earth than when it's waxing or waning (foreground), and it appears much smaller and less bright. This is why the naked eye gets no hint of its phases.

of astronomical theories and observations. The changeable shapes of shadow, interpreted properly, allowed him to glimpse the orbit that Venus makes around the sun. In a famous passage of the *Saggiatore (The Assayer)*, Galileo says that science is written in the book of the universe, which lies open before our eyes. You must only learn to read it; you must learn the language of the universe, mathematics: "The characters are triangles, circles, and other geometric figures, without which tools it is impossible to humanly understand the words; without them you wander in vain through a dark labyrinth."

The labyrinth is dark, but the shadows surrounding us can have a shape that guides us toward knowledge.

The phases of Venus as drawn by Kepler; Epitome of Copernican Astronomy, *part III*

Mercury Unveiled by Shadow

Other skies. After two days of uninterrupted rain, at nine in the morning on November 8, 1631, the sun pierced the clouds of Paris. A ray of sunlight filtered into a dark room and passed through a telescope that projected it onto a page. The image of the sun was about 8 inches in diameter. Pierre Gassendi (1592–1655), an atypical philosopher and priest—for one thing, he didn't much like Aristotle—observed a dark dot on the surface of the sun. Could it be a sunspot? Unsure, he noted its position. The sun came and went. In the rare intervals of good visibility, the dot moved—but it moved much faster than a sunspot would. Located on the surface of the sun, sunspots take twenty-six days to complete a rotation; but in just a few hours this dot crossed the whole circle of the sun. So was this really the expected event? The event that was predicted by Kepler and observed only by Gassendi and by a few other lucky souls who had the bright idea of using a telescope instead of a simple camera obscura to project the sun's image? The dark spot was the silhouette of Mercury passing in front of the sun. And the fact that it was just a little dot also explains why only a few people considered it necessary to use a telescope. Even Gassendi didn't believe his eyes: Mercury had a diameter of only 20" of an arc, one-seventh the size it was usually thought to be. Kepler, for example, had estimated that during its transit Mercury would have measured an apparent 2.5' in diameter (one-twelfth the diameter of the sun); but his idea of the planets' sizes was based on an a priori predilection for harmony among celestial numbers. When he drew the scientific community's attention to the need to observe the transit of Mercury, his overestimation of Mercury's size meant that almost no one was adequately prepared to do it. (Kepler himself died a year before the transit; he was spared the disappointment.)

Remus Quietanus, one of the fortunate few who observed Mercury in front of the sun, did guess Mercury's size, because his speculation on the harmony of cosmic sizes led to results different from Kepler's. His theory was that all the planets would look the same size when seen from the sun. We might even dream of a magical alignment: from the sun we could see the mother of all eclipses, with each planet hidden tidily by all the other smaller planets in front—Jupiter hidden by Mars, Mars by Earth, and so on—so that only Mercury was visible. Remus was very happy when the size confirmed his strange theory.

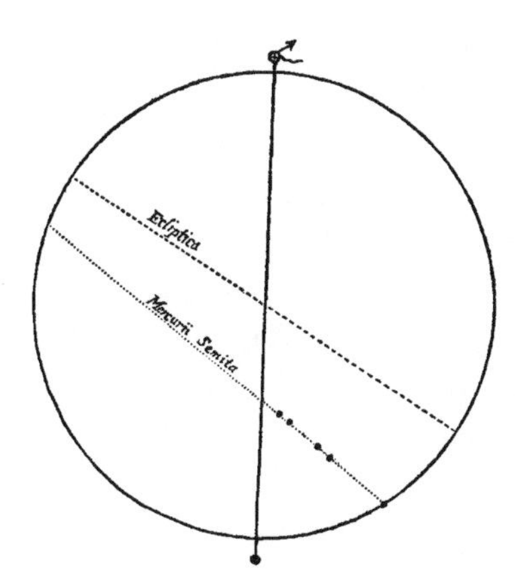

The little dot is Mercury, seen against the sun by Gassendi on November 8, 1631.
This is the size that Mercury was expected to be: ●

But why was Kepler so wrong about Mercury's size? We see the planets in the night sky because they reflect the light of the sun, and we cannot distinguish them at a glance from the stars except by their brightness. Galileo's telescope stripped the planets of their halos, but the stars kept their luminous crown. So the planets were seen as things, shadowy boulders, while the stars were still lights. But although Venus, whose phases were visible, could be clearly distinguished with the telescopes of Galileo's time, that was not the case for Mercury. At that time the distinction had not been made between the size and the luminosity of heavenly bodies. Even today the "magnitude" of a star is not its size, but its apparent luminosity; this is part of the classification that goes back to Hipparchus and the Greek astronomers. So Mercury's size was easily overestimated on the basis of its very bright appearance. Gassendi's observations were the first to supply a reliable estimate. In Mercury's case the shadow indicates the dimensions of the planet more faithfully than light does.

Venus's Revenge

Venus too appeared larger than it was, and it couldn't avoid being drastically reduced; but that came later. In 1639 the young and unlucky self-taught astronomer Jeremiah Horrocks and his correspondent William Crabtree were the only ones to observe the extremely rare passage of Venus in front of the sun, and to ascertain the planet's small size. The former died a few months later, at just twenty-three, on the eve of a journey in which he was to deliver his astronomical report to Crabtree. Crabtree then appears to have been killed in the Civil War in 1644. So Horrocks's manuscript wasn't known for some time; not until another passage of Venus in 1769 was the measurement confirmed. In the old order of the cosmos, Venus was a star, and it was the brightest star. Galileo considered it an imitator of the moon; Horrocks turned it into a minor planet. The goddess of love was losing the aura that had for so many millennia made her an object of admiration; her vanity would not allow her to tolerate this.

Punishments

As soon as there were strong enough telescopes, and people began drawing the first lunar maps, a problem arose: what should lunar craters be called? Among the various proposals, the one that still holds today was that of the Jesuit Giovanni Battista Riccioli (1598–1671), who labeled the craters with names of illustrious scientists on a map drawn by the optics researcher Francesco Maria Grimaldi (1618–1663) and published in 1651. (Incidentally, the Jesuit Grimaldi was a fairly important man of shadows: he discovered light diffraction by studying shadows.) Riccioli passionately defended the theory of the earth's immobility, and so he relegated the Copernicans to the Sea of Tempests (which laps at the Peninsula of Delirium). A brighter configuration than the others was named for Galileo but was later revealed to be just a long whitish scar, perhaps from the impact of a comet.

Galileo's name now graces two silly little craters nearby. On the map, at least, this looks like a posthumous victory of the Jesuits over Galileo. The craters stand as a strange reminder of an era in which scientific debate got tangled up with the subtleties of professional metaphysics and sometimes stooped to making showy analogies. Galileo ended up here, hidden in an obscure lunar sea, in the heart of one of those spots that were so dear to Plutarch and that threw astronomy off course. But maybe it's not such an ignominious ending for the geometer of shadows.

XIII *Maybe Saturn Devoured His Own Children?*

Let's take a step back. On August 4, 1610, Galileo communicated to Kepler another one of his important early astronomical discoveries, by writing to Giuliano de' Medici, the Tuscan ambassador in Prague, in a coded text:

Smaismrmilmepoetaleumibunenugttaurias

Faced with this anagram, Kepler—who loved to "gnaw on bones and hard crusts of bread"—didn't give up. After much effort he extracted the pidgin Latin phrase *Salve umbistineum geminatum Martia proles* (Hail, O flaming twin, son of Mars); for a short time he believed that the Red Planet had a moon. (The idea that Mars had a satellite was already in the air, for reasons of cosmic harmony: Mars lies between Earth—which has a moon—and Jupiter, around which four moons were seen. Mars does in fact have two satellites, two great rocks that, however, were discovered only in 1877.) Not completely convinced of his solution, Kepler published the anagram in his *Narration* on Jupiter in which he confirmed Galileo's observations about the little "Medici" satellites, and he expressed his hope that someone would decipher the anagram better than he, or that Galileo would reveal the secret. The solution came in a letter of November 13 from Galileo that read *Altissimum planetam ter-*

migeminum observavi (I have observed the most distant planet to have a triple form). The observation was imprecise: Galileo's telescope didn't have sufficient definition to reveal Saturn's shape, and Galileo thought he glimpsed two enormous moons standing immobile at the planet's sides.

It took almost half a century—and a remarkable improvement in telescopes, as well as the contributions of many mathematical minds—before the strange picture of Saturn could be understood. Christian Huygens (1629–1695) announced the solution in 1656 in his own way:

Aaaaaaacccccdeeeeeghiiiiiiillllmmnnnnnnnnnooooppqrrstttttuuuuu

If you don't have Kepler's determination, there's no point trying to decipher the anagram. Let's look first at Galileo's mysterious triple planet.

We can use a simple device to see what went wrong with Galileo's first observations. Below at left I've drawn some shapes that look like Saturn when its ring is tilted so that, from the earth, we see more of the ring's surface at the edges than in the middle. At right I have "softened" the pictures with computer manipulation. In 1610, Galileo must have seen something similar to these unfocused views.

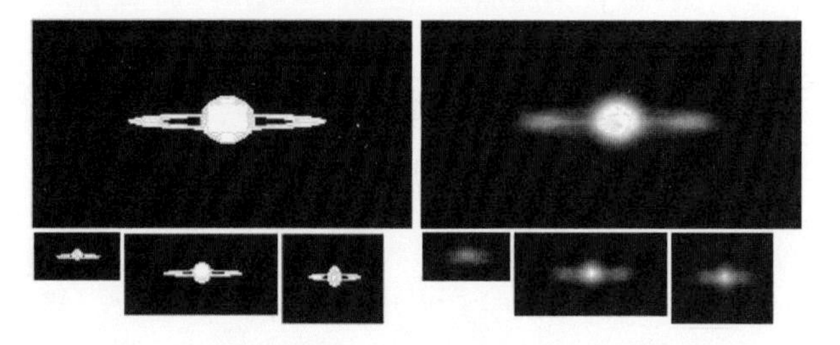

Four drawings of Saturn's shape, and (at right) the same pictures in a slightly unfocused version. The right-hand pictures—which give an idea of what Galileo may have seen—don't make it easy to see that there is a ring there.

The two large satellites were quite odd anyway, in that they stayed still and didn't circle their planet. Furthermore, toward the end of 1612, Galileo had some doubts about Saturn's appearance. After two years during which he had always seen it as a large body flanked by two minor bodies, he now saw it standing alone, like all the other planets. "I found it solitary, without the assistance of the usual stars, and all in all perfectly round and bounded like Jupiter. . . . Maybe the two minor stars have been consumed, like sunspots? Maybe they disappeared and suddenly ran off? Maybe Saturn devoured his own children?"

At this point Galileo had various hypotheses to explain Saturn's new look, but it's interesting that he doubted his eyepiece. "Was it illusion or fraud, the appearance with which the crystals fooled me and others for so long?" Here he puts his finger on the problem. In the coming years, determining Saturn's shape would be an obstacle for all telescope builders. The competitors in this race were skilled artisans and ambitious scientists. A plethora of hypotheses filled astronomic correspondence: Saturn is a planet with two handles! Saturn is an oval with two holes! Saturn has an elliptical corona that sticks to it while the planet oscillates! Galileo himself changed his mind in 1616 and drew a kind of ovoid configuration, located *behind* the planet this time. When Huygens published his *System of Saturn* in 1669, he included a mocking graphic summary: in one illustration he recorded all the distorted pictures of Saturn produced by illustrious astronomers and astrophiles before him.

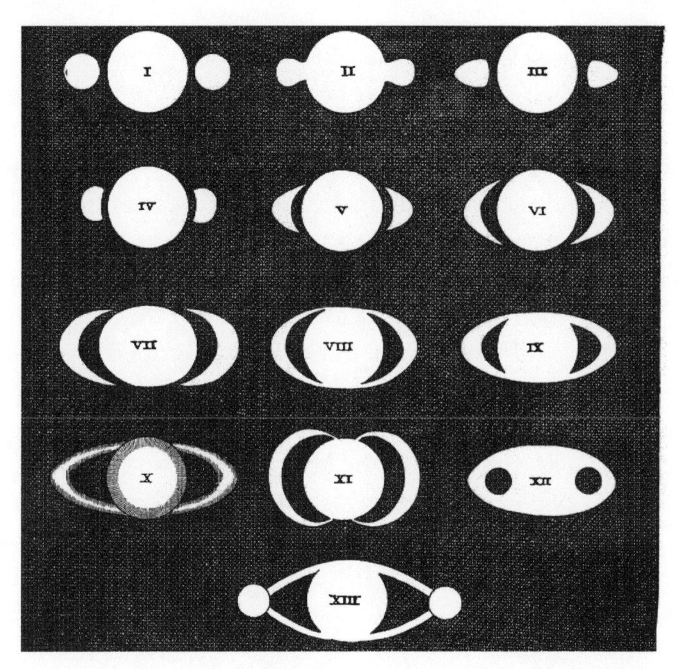

The Museum of the Horrors of Saturn: observations by Galileo (I), Scheiner (II), Riccioli (III, VIII, and IX), Hevelius (IV–VII), Fontana (XI, XIII), and Gassendi (XII). Try standing 5 or 6 meters away and, in fairness, you'll see that it's not easy to distinguish among these shapes. Number X is a Saturn by Divini, who drew shadows that he didn't see and which don't even exist.

Drawing number X is interesting for our discussion. It's different from the others because it shows bizarre shadows. Huygens had copied it from an advertising card published in 1649 by Eustachio Divini, a gifted telescope-maker without much skill in astronomy. The card promised all kinds of rewards to his potential clients, investigators of the cosmos; thanks to Divini's telescopes, they would see marvelous views of the sky, including the waxing Venus and the equally popular moon, with all its mountains and pockmarks; it also promised views of a magnificent Saturn framed by an ovoid disk or ring.

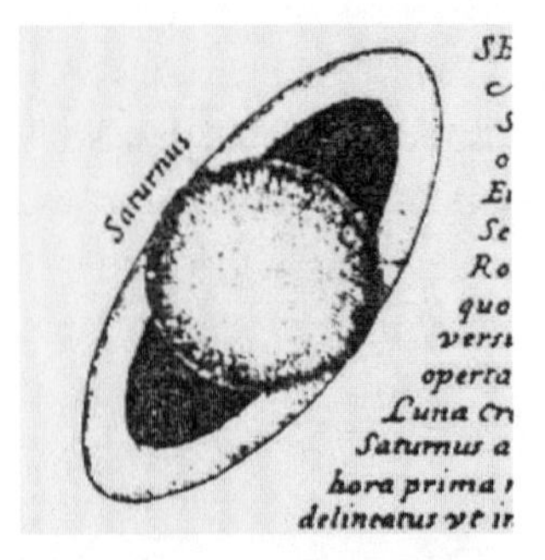

A corner of the advertising card for Eustachio Divini's telescopes: "Buy Divini's Telescopes, and You'll See This and So Much More!"

But the shadows in drawing number X are strange, and they certainly don't correspond to solar illumination. (A tiny sun would have to be jammed up against Saturn.) Huygens declared the shadows false. Divini was obliged to admit that he added them to show the shape of the planet better. The two men began to debate the quality of their respective telescopes and the nature of Saturn. But despite his criticism of the shadows, Huygens appreciated Divini's drawing because, unlike all the other monstrous Saturns, this one came close to his hypothesis about the planet, which was concealed in the letters of his 1656 anagram: *Annulo cingitur tenui, plano, nusquam cohaerente, ad eclipticam inclinato* (It is surrounded by a thin, flat ring, nowhere touching it, and inclined to the ecliptic). Divini saw the ring clearly, but he didn't understand its position in respect to Saturn. Huygens's idea was simple: Let's use shadows to verify the ring hypothesis. If the ring is inclined in respect to the apparent trajectory of the sun in Saturn's sky, it should cast a shadow on the planet—and the planet in turn should shade the ring. Saturn is so far from the earth that these shadows will be minimal; but it should be possible to see them anyway.

The strange thing is that in 1656 Huygens really saw shadows on Saturn—but he didn't understand that they were shadows! His picture of a 1656 view of a "solitary" Saturn (without its usual accompanying handles) shows a dark band crossing its middle.

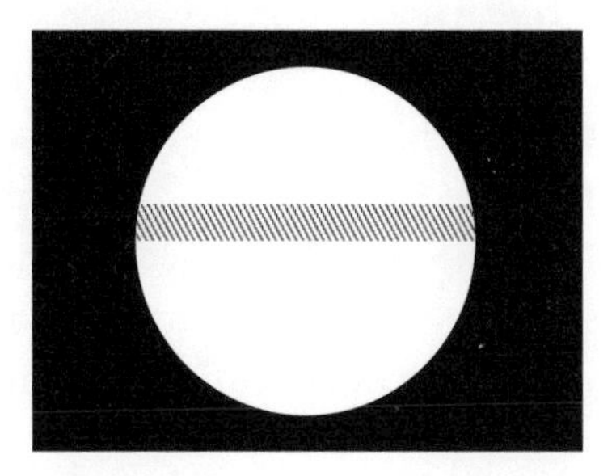

Huygens saw a shadow but didn't understand that it was a shadow.

Huygens thought this was the thickness of the ring, seen edgewise. But in reality the ring is very flat (200 meters on average; seen from Earth, it's like a thin sheet with no depth), and the dark band was its shadow, cast from "below."

Many people saw this band in 1655 and 1656 without understanding that it was a shadow. Huygens was so satisfied with his hypothesis that the ring was somewhat thick that he even falsified a drawing—he sketched a shadow that wasn't there, and he failed to show what should have been there. This sketch, representing Saturn in 1657, suggested a nonexistent band again and did not show the shadow that should have appeared above the ring. (We know that the shadow should have appeared in that position because we can work backward to calculate the relative positions at that time of Saturn, the sun, and the earth.)

Saturn seen from the Voyager *on October 30, 1980. The shadow cast by the disk onto the planet is more than 6,000 miles thick. The ring itself is much thinner. Note the enormous shadow cast, in turn, by the planet onto the ring, barely visible from Earth.*

The Hunt for a Shadow (No Blow Is Too Low)

In 1656 the hunt for a shadow—the right shadow—began. It was first seen on August 20, 1660, from Florence: it was a spot that interrupted the continuity of the ring. Huygens observed the shadows several times between 1664 and 1693. But the sharpest views—definitively confirming the hypothesis of the rings—were drawn by Giuseppe Campani, a telescope-maker who in 1664 launched his product with an aggressive marketing plan: he challenged and quashed Divini, his competitor, in a battle of long-distance observation, with each man using his own instruments. (Campani's telescopes routed the competition: practically every important phenomenon registered in the sky in the latter half of the 1600s was observed through one of Campani's pieces.) Campani was certain he would prevail; he had been certain ever since 1663, when he looked for Saturn and saw the planet shadowing the ring and the ring shadowing the planet simultaneously.

In the meantime, Campani had a further victory when Cassini in Paris used a Campani telescope and glimpsed the microshadows made by Jupiter's satellites as they passed over their planet. Campani included these shadows too in his own advertising card:

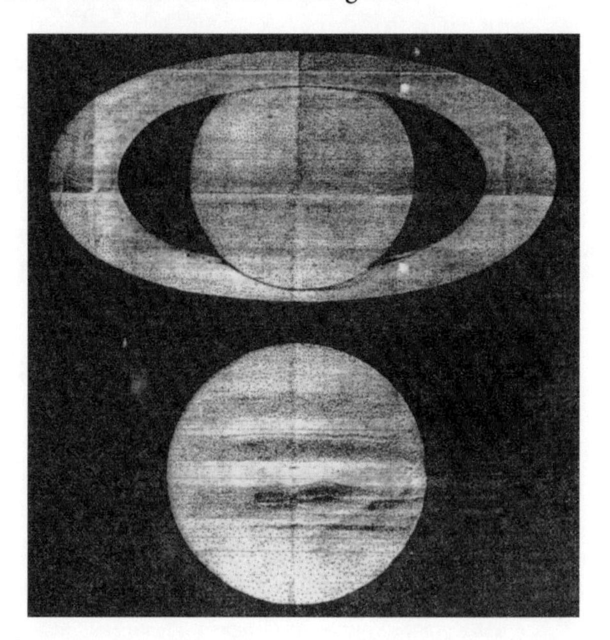

The shadows on the rings of Saturn and the little shadows of Jupiter's satellites, seen in Giuseppe Campani's drawing. This was just a publicity flyer, but scientific illustration was growing more important everywhere.

Campani's ad was less beautiful than Divini's, but Campani knew how to convince the connoisseurs. Cassini wrote with amazement that he never would have believed he could see Jupiter's little shadows, small as they were. By this point a telescope's power was measured by its ability to show the farthest shadows of the solar system. And these were the most distant shadows ever seen by the human eye.

What a Saturnian Sees

Saturn was Huygens's great passion. We can see this in his posthumous work of popular science, the *Cosmotheoros,* in which Huygens tells how the inhabitants of other planets might write astronomy from their own point of view. (This mental experiment is tremendously useful for teaching. Kepler had tried something similar with the moon, to promote the Copernican system. In his version the inhabitants of the moon are absolutely convinced that the earth is not immobile, because from where they are, they can *see* it spinning.)

Saturn's astronomy is made spectacular by the moons that circle around it. There are eighteen of them, but even the five that were known in Huygens's time must have had a strong impact. The big deal, though, is the separate issue of the ring, which produces breathtaking shadow displays:

> Around the poles is a zone . . . where the inhabitants (unless maybe the cold makes these zones uninhabitable?) can never see the ring. At all other points of the surface it can be seen continuously for fourteen years and nine months, half a year for the Saturnites. During the other half of the year the ring is hidden. Those who live in the vast zone between the polar circle . . . and the equator . . . that lies beneath the ring can see, in the middle of the night—for the whole time that the Sun illuminates the side of the ring that faces them . . . a bright arc that rises from the two sides of the horizon but is interrupted in the middle by the shadow of Saturn . . . often all the way to the other end. But after midnight the shadow moves bit by bit, toward the right for a spectator in the boreal hemisphere and toward the left for a spectator in the opposite hemisphere. And this shadow disappears in the morning, but the appearance of an arc remains. . . .
>
> When the globe's shadow is cast on the G–H part of the ring,

> the ring's shadow covers a zone of the globe toward P–F that otherwise would receive light from the sun. So there is a zone P–Y–E–F, sometimes wide and sometimes narrow, whose inhabitants are denied the sight of both the sun and the ring for long periods, while the ring simultaneously hides some of the stars from them. And this must have the effect of a miracle when they find themselves in deep darkness without seeing what causes it.

The picture in the original edition shows the ring's shadow rising along the planet's side from point F up to point P. When the *Cosmotheoros* was translated into English in 1698, the engraver intervened, "correcting" what he thought was a defect in the original edition: he eliminated the shadow that the ring casts on Saturn (from point P to point F).

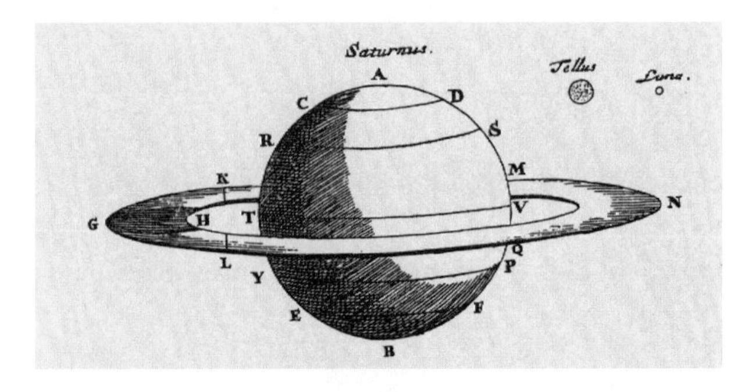

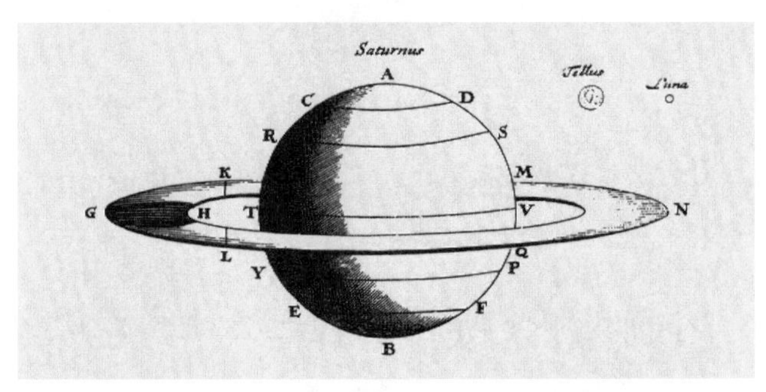

Saturn, in Huygens's Cosmotheoros. *Note that the ring's shadow on the planet (in the zone P–F) has disappeared from this picture included in the English edition of 1698, probably because a zealous engraver considered the original edition to be wrong.*

If there were no ring, the engraver would be right. Clearly, the artist was conversant with the rules of perspective drawing of shadows.

And his knowledge of the laws of drafting misled him: the shape of the shadow in the original engraving must have seemed incongruous to him.

So far, this has been a sketch of some major events of the seventeenth century. Chasing after light—trying not to lose the traces of shadow that heavenly bodies imprint on light—the telescope dulled the shine of the moon and left it a pockmarked rock, unmasked Venus and draped it in shadows. But the telescope rehabilitated Saturn. The slow-moving old man of classical mythology, the cruel father who devours his own children, turned out to be a lovable cosmic plaything.

XIV *The Speed of Shadow*

If Saturn is a plaything, Jupiter is a scientific tool. More precisely, Jupiter is a clock.

I like to imagine that I could create a shadow even bigger than Jupiter's shadow. When I point my cat's flashlight toy into the night sky, the mouse shadow stretches into the cosmos infinitely as the light reaches and spreads: the bulb is smaller than the mouse profile. All the shadows created by the sun, though, no matter how enormous, are finite in size. The sun is too big a light source compared to the planets, so all the planets' shadow cones end in a point.

But those shadows are still powerful. Saturn casts shadows longer than any other body circling the sun: they average a bit more than 80 million miles in length (this is almost the distance between the earth and the sun). Even Jupiter's shadow, not the biggest of the lot, stretches almost 56 million miles. In its dark cone four satellites hide one after the other, each as big as our moon, and quickly reemerge into the light after a short run in the shadow. Galileo sighted them first. As we saw, he reported the discovery in the *Starry Messenger;* by dedicating the discovery to his patron, he won himself honors and a coveted sinecure. But his relationship with the Medici planets went on for another two years, until 1612, when he was finally able to determine their periods (the cycles of their revolutions around Jupiter) and turn Jupiter into a clock.

Galileo's perseverance with this problem was impressive. His notebooks recording his efforts tell a story of his difficult progress, of intuitions that turned out to be wrong, of observations that were hard to fit with one another. But the challenge wasn't only intellectual. Galileo was betting that the glory he had won by unveiling the planets would be completely eclipsed by the discovery of their cycles. He thought this because Jupiter's planets immediately appeared to be the solution that everyone sought to the problem of longitude.

While the problem of latitude (how far north or south is this point from that other point?) was fairly easy to solve, the problem of longitude (how far east or west is this point from that other point?) had long been a serious puzzle.

What's the difference between the two problems? The difference is that the earth spins. The axis of the globe is a fixed reference point that pierces the planet at the poles. If you use a celestial reference point to determine the position of a certain place, then the earth's rotation is not a problem—indeed, you can exploit the very fact that the earth's axis stays fixed (over a sufficiently long period). If you know, for example, that the North Star is more or less directly above the North Pole, you then know that it's a fixed reference point in the sky. The angular distance from the pole, and thus the height of the North Star over the horizon, is one measure of your latitude. (At the equator, for example, the North Star is right on the horizon.) But there is no East Pole or West Pole for measuring longitude; the direction with which you measure longitude keeps "slipping away" as the earth spins.

We can look at the problem another way. When we calculate the angular distance between two points on the earth, whether latitude or longitude, we need to *exchange information* between these two points. The information measures the incline of the ground compared to certain reference points. Eratosthenes, for example, gathered this information by studying the shadows cast by a stylus; his reference point was the direction of the sun's rays. Let's consider latitude. We know that shadows get longer the closer we come to the poles. But we notice immediately that we have to measure the shadows at the proper moment of the day: if an astronomer in Alexandria observes his stylus *in the morning* he'll see a *long* shadow; if an astronomer in Rome observes his own stylus at *noon* he'll see a *short* shadow; and when comparing the two, we might conclude erroneously that Rome is *south* of Alexandria. The two astronomers must agree on a certain observation time during the day.

If they measure the shadows at the same local time, they can factor in the difference between the shadows' lengths caused by the day moving forward—or rather, by the rotation of the earth. It's easy to determine the right moment, because every point on the earth's surface moves along the line of a parallel, and at a certain time the midday line (the moment when the sun is highest and the shadows are shortest) will reach Alexandria; sometime afterward it will reach Rome. Here we are helped by the fact that the earth spins, because this allows Alexandria and Rome to communicate, transmitting information about midday. All they have to do is compare the shortest Alexandria shadow with the shortest Rome shadow.

We have the opposite problem when dealing with longitude, because the difference between the local times *is* the difference in longitude, and it's impossible to separate shadow's two aspects, as we did with latitude. Here the fact that the earth spins is an obstacle. How can we overcome it? If only we could send an instant message to Rome saying *it's noon in Alexandria right now.* Knowing that it's noon in Alexandria, all we have to do is determine the local time in Rome in order to calculate how long until it's noon here. The earth turns 15° per hour, so the difference between the time in Rome and the time in Alexandria, multiplied by 15°, gives us the difference in longitude between Alexandria and Rome (give or take a few minutes).

But how can this information be transmitted immediately from Alexandria to Rome? Nowadays we could simply phone; this solution wasn't available in the preelectric era. (The introduction of rapid communications, with telegraphy and telephones, was the determining factor in the creation of a system of universal temporal reference. Even a less rapid information transfer, like a letter sent by train from one coast of the United States to the other, wreaked havoc on the habit of using local noon time in the small-town stations along the way. Rapid communication not only solved the problem of longitude once and for all; it also demanded the construction of a time grid that put the world on a syncopated rhythm. When people traveled from Alexandria to Rome on foot and by boat, they reset their watches by a few minutes each day; when we make the trip by plane it makes more sense to set our watches ahead just once, by a whole hour.)

To immediately exchange information about local time, Rome and Alexandria need to *see the same event at the same moment.* Which means that they have to look at some timekeeper that isn't affected by the earth's rotation while it's marking time. A sufficiently precise and trustworthy clock set on Alexandria time, traveling from Alexandria to Rome, would permit an observer in Rome to see what was the local time in Alexandria as well as the local time in Rome. But equally helpful would be an astronomical event that was independent of the earth's rotation and visible simultaneously in Rome and in Alexandria.

Appointments with Shadow

This is where cosmic shadows come into play, because eclipses are good clock events. A lunar eclipse is visible at the same time in the same way from every point on the earth where it's nighttime. It is as if you could read a clock mounted like a gemstone in the sky. All you have to do is observe a lunar eclipse from Alexandria and from Rome and then compare the moment of the eclipse with the local time in each place: you can thus find the difference in longitude between the two points. Al-Biruni did just this; he wrote, "If we know ahead of time about a lunar eclipse and we wish to determine the difference in longitude between two cities, we can make an agreement beforehand with people in the two cities who are able to accurately measure the [local] time with instruments, to obtain the precise moment of the beginning and of the end of the eclipse." But lunar eclipses are not frequent enough; and coordinating a network of astronomers who live in far-flung cities is quite an organizational burden.

Once people could see them in the sky, the moons of Jupiter turned out to be much easier to use. First of all, there are lots of them. Furthermore, their cycles around the planet are shorter than those of our moon around Earth. Their plane of revolution around Jupiter is only slightly inclined, compared to the direction of the sun's rays. And, most important, Jupiter makes a big shadow. The upshot is that eclipses follow one right after the other. Jupiter's system is a great big sundial, in which the sun and the stylus (Jupiter) stay still, and the satellites in their rapid succession of rotations around Jupiter furnish the mobile dial on which the planet's shadow marks the hours. And even though the clock's ticks and tocks are not all equal—because the cycles overlap and some of the ticking runs together—these shadow appointments follow a fairly precise schedule.

So Galileo spent a couple of years struggling with the cycles of Jupiter's satellites, in the face of his peremptory colleague Kepler's unhelpful dismissal of the task as practically impossible. But finally

Galileo did it. Once he got the desired result, he redoubled his efforts to make it easy and practical to use, especially for the navy. Two inventions remain as evidence of this effort. The jovilabe, or astrolabe of Jupiter (Jove), was a curious brass instrument for calculating the positions of Jupiter's satellites relative to phenomena visible from Earth: the satellites' conjunction with and separation from the planet, the conjunction of two satellites, and a satellite's moving into Jupiter's shadow or out of it. The celatone was a sinister-looking metal mask with a built-in telescope; with your free eye you could look at the planet, and then you could observe it in detail with the telescope mounted over the other eye. Neither of these instruments was particularly easy to use, and Galileo failed in his efforts to get recognition (and a reward)—from the various nations that had their own naval fleets—for his solution to the problem of longitude.

Although it didn't garner Galileo any of the prizes that he hoped for, and although it didn't help sailors to find their position at sea (which not only lacks reference points but is constantly in motion), still the connection between Jupiter's shadow and the motion of its moons did solve the problem of *terrestrial* longitude and permitted the creation of far more accurate maps than had ever been made before. It is strangely ironic. Almost two millennia earlier, shadows on Earth had already solved the problem of latitude, but we had to look for a shadow in the sky in order to deal with the problem of longitude; and while terrestrial shadows are small—only about as big as the human artifacts that measure them—Jupiter's shadow is so large that it can swallow the whole earth.

Time Travel, the Black Day, and Two Modern Discoveries About Shadow

Solar eclipses allow us to calculate spatial differences even more accurately than do lunar eclipses. In New York on January 24, 1925, if you walked north across 96th Street, you could move out of the area of total eclipse and see a thin sliver of sun. On a larger scale, this phenomenon permits us to gauge the difference between two locations where an eclipse is visible. (In the 1700s the method greatly interested the administrators of the North American colonies, who were struggling to draw maps for such a large continent.)

Recent discoveries suggest that, over a long period, eclipses also allow us to discern more subtle differences, minimal temporal discrepancies. If the Eclipse-Tourists Club were to get together with the Time-Travelers Club, they could organize trips back to the great solar eclipses of the past. A journey through time is always a journey through space-time, and before you start the engine on the time machine, it's important to carefully calculate the moment and the area where you wish to view the eclipse. To do the calculation you just have to run backward through the same steps that allow us to predict future eclipses. But oddly, the simple calculation is not enough: a trip back in time to catch the eclipse of 136 B.C. in Majorca would be disappointing. Once they landed on the island, the tourists wouldn't see an eclipse! That's because the organizers would have depended on astronomers and not historians, and so failed to read the Babylonian tablet in the British Museum that refers to that date. "At 24 degrees after dawn—solar eclipse. At the beginning toward the southwest, at 18 degrees in the morning, it became total. Venus, Mercury, and the usual stars were visible. Jupiter and Mars, which were in their period of invisibility, became visible. The sun sent out its shadow from south-west to north-east." The Majorca eclipse is actually happening in Babylonia! This wouldn't be the only wrong trip. Richard Stephenson and Leslie Morrison analyzed seven hundred reports of historic eclipses and observed that each one actually happened farther east than the locations pinpointed by modern calculations. Because the earth rotates from west to east, "happened farther east" actually means "happened earlier than expected." Why in advance? The calculations had been done with the premise that the earth rotates at a constant speed; Stephenson and Morrison's conclusion is that the earth's rotations are slowing down. Not by much: today's day is approximately one-twentieth of a second longer than a day 2,500 years ago—although it's not clear whether the slowdown has been constant or whether it's just one phase in a much larger cycle.

Another interesting bit of data from reports of past eclipses concerns the size of the sun. The first careful survey of an eclipse's path exists thanks to the efforts of Edmund Halley (1656–1742), the astronomer who gave his name to the famous comet. Because he didn't have the funds to research and gather data, he published a flyer with a disturbing and curiosity-provoking title: "The Black Day or a Prospect of Doomsday exemplified in the great and terrible Eclipse which will happen on the 22nd of April 1715." The sheet, showing a map with a calculation of the path of totality, was widely read, and the letters that Halley received in the following months showed that the moon's shadow actually passed a bit farther south than expected. The corrected map that Halley derived from these reports was used in 1980 by David Dunham, who, with the usual system of reverse calculation of the eclipse's path, saw that the shadow was smaller than one would expect from today's data—which leads to the hypothesis that in the meantime the sun has shrunk slightly.

The Speed of Shadow

Its satellites' eclipses serve as the ticking of the great clock that is Jupiter. When a satellite moves into the shadow, it's as if it makes a "tick"; when it emerges, that's the "tock." (Depending on the earth's position, we see only the tick or only the tock, but we know how much time passes between one and the other.) It would be hard to imagine a more stable clock. And yet in September 1676, Ole Rømer, a Danish astronomer working in the Paris observatory, made a staggering announcement: the eclipse of Jupiter's first satellite (known as Io today; of the satellites then known, it is the closest to the planet), expected on November 9, would appear exactly ten minutes later than the time predicted by the astronomic tables. What was happening? Was Jupiter's system not so precise after all? And how did Rømer figure this out?

Rømer had watched Jupiter for several years, measuring the slow tick-tock. He realized that the ticking slows down until the moment when Jupiter and Earth are on opposite sides of the sun, and then it speeds up again until the orbiting Earth slips in again between Jupiter and the sun. Basically, when Earth, in its solar revolution, moves toward Jupiter, the clock speeds up, and when Earth moves away, the clock slows down. But how does Earth's movement slow the ticking of Io around Jupiter?

The truth is that Earth does nothing. The explanation of the phenomenon is one of the greatest shadow discoveries.

When Jupiter and Earth are on opposite sides of the sun, they are as far apart as they can be. When they are on the same side, they're pretty close. The difference between the two distances corresponds to the diameter of the terrestrial orbit. *We know that Io has been eclipsed because we can see it. The light acts as the messenger for this event. The ticking of the satellite seems to slow down when Jupiter is farther away from us, but that's only because the light takes longer to reach us now than it does when Jupiter is nearby.* The eclipses of Jupiter's satellites don't

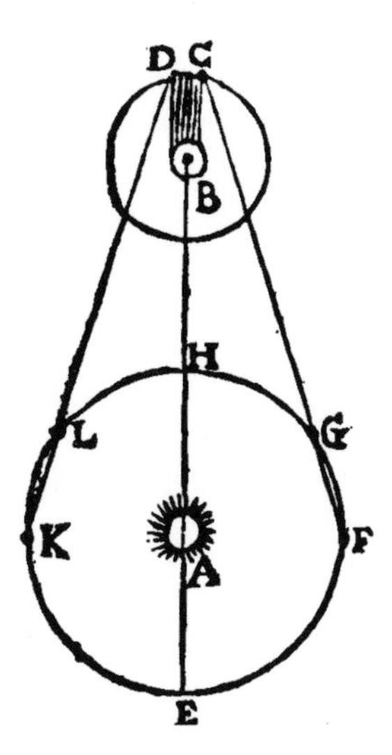

Going from L to K, Earth in its solar cycle sees Jupiter's first satellite slowing its rhythm of eclipses. But going from F to G, Earth sees the rhythm speed up.

slow down; it's only the messenger light that—though it travels very fast—is not infinitely quick. Rømer's discovery was that *the speed of light is finite.*

This was the final salvo of fireworks in the astronomers' wars during the century of shadows. A curious Fate wanted the speed of light to be unveiled by a slow shadow.

Part Four SHADOW VISIONS

(Curtain rises)

Plato and His Shadow

In the olive groves near the port of Piraeus. Skia, Plato's shadow, stretches along the ground and climbs the stone wall bordering the path. A lizard, chilled as Skia passes over it, skitters away.

PLATO: A whole new world. And yet . . .

SKIA: You're still doubtful? You're still afraid? Weren't the heroic feats of Galileo and all the other guys enough to convince you?

PLATO: They certainly were remarkable. But you saw how much trouble people have dealing with you. Look at that illustrator, for example, the poor guy who forgot about the shadow of Saturn's ring.

SKIA: You really don't get it! He made that mistake because he's *too* good.

PLATO: Hunh? Don't you think you're letting your paradoxes run away with you? You're grotesque.

SKIA: Let me explain. It's true that the illustrator made an unforgivable mistake when he disregarded the rings. He concentrated only on the sphere. But how did it happen? The shadow of the original drawing struck him as odd: that *couldn't* be a sphere's shadow. So he corrected it.

PLATO: So?

SKIA: It felt natural for him to fix the drawing, which means that in his day you couldn't be an artist if you hadn't mastered shadows.

PLATO: Who'd have thought so? But why is all this important?

SKIA: Just consider Galileo's challenge: Give me any shadow and I'll reconstruct a shape. Galileo could win such a bet because painters had solved the opposite problem for him: Give us a shape and we'll construct its shadow.

PLATO: How did they manage to do that?

SKIA: Well, let's see.

XV *The Shadow Line and the Shadowy Rays*

God loves only the elect. . . . Not everyone can be saved: this can be seen in the universal harmony of things—painting that lives because of shadow, and consonance because of dissonance.

—Leibniz, *The Philosopher's Confession*

If you were having your profile drawn, whom would you trust more—the person who sketches you freehand or the one who traces the outline of your shadow?

Philosophy and astronomy are born of shadows. Pliny would have liked to add painting to this list. Of course, his story tells us, the origins of painting are obscure: the Egyptians claim to have a six-thousand-year head start, but that's just an empty boast. The Greeks have a better tale: they date the beginning of pictorial art back to the moment when the profile of someone's shadow was traced on a wall. The same myth is linked to the birth of sculpture. Here three characters enter the stage: Butades, a vase-maker from Sicyon working in Corinth; his daughter; and her lover. Before the lover left for a distant land, the girl traced his shadow's outline on the wall. The next day her father carved the shape out as a bas-relief. Painting and sculpture were born when a shadow was captured on a wall by the work of a human hand.

Maybe in Pliny's time (the first century A.D.) these myths sought to explain the ancient images of Egyptian and Greek painting, the black silhouettes populating frescoes and ceramics, which in Greece were referred to as *skiagraphia*—shadow painting. These ancient pictures walk in profile, and like shadows, they are monochrome and lack internal detail. But beyond the historical explanation of the myth, there is a reason why people found it so easy to believe that painting was born from the tracing of a shadow. Myths spread only if they conjure up powerful images, not because they echo a historical event. What grabs the imagination in Pliny's story is the idea that the painter is only a minor character: *it's the shadow that comes into the limelight as it does most of the work.* The casting of the shadow is a natural process that follows geometric laws, and that's why we can trust the outcome. In other words, there is a direct line from the model to its reproduction, a line that doesn't even pass through the fallible mind, and the uncertain hand, of the artist.

The Taboos of Shadow

There is indeed some reason to mistrust the demiurges of images. Although the argument is that painting was born from shadow, there is no other intellectual discipline in which the battle with shadows has been more intense than in painting. The most obvious sign of this struggle is the almost complete eradication of visible shadows from the pictures of every culture. In just a few, rare representations, the shadow itself is the subject of the story, and the painter could not afford to ignore it; but in the overwhelming majority of cases people have preferred to avoid it. Shadows almost never appear in pictures from non-Western cultures; and in Western art before the Renaissance, shadows were in and out of fashion. Even in recent times artists have had an ambiguous and unresolved relationship with shadows. In tenebrous Mannerist and Baroque paintings, shadows of people and things get lost in the darkness of poorly lit rooms.

We could suggest various possible reasons for the absence of painted shadows, which starts to look a lot like a cultural taboo. One deep explanation could be metaphysical: it could be tied to shadow's strange properties. Shadows are troubling things—dangerous hiding places for highwaymen; they are bothersome replicants of human figures. After all, they crowd the canvas and distract the viewer. Another hypothesis is that we rarely pay any attention to shadows, and our paintings simply reflect this inattentiveness.

But maybe the real reason lies elsewhere—maybe it's much more banal: it's hard to render a shadow well. Shadows have only recently started to appear in animated cartoons—only since it became possible to generate them automatically. (And even so, many digital movie studios employ large "shader teams.") Architects and decorators alike appreciate the services of the many computer-graphics programs that build shadows right in. Two things are at issue here: light and geometry. As the great psychologist Hermann von Helmholtz (1821–1894) noted, a picture on an opaque surface—such as a painting—uses only a very limited range of values for brightness and darkness. The whitest patch on a painting (a picture of the sun at midday) reflects not much more light than the darkest patch (the depths of a cave). If you squint to look at a landscape painting, you'll see that the light in the image merges together into an incredibly dark splotch. In reality, there's an enormous difference between the light of the sun and the light reflected back from a shadowy area; it's surprising how much the brain can reconstruct when it looks at a painting. Just consider your own shadow on a sunny day and how much it can reduce the amount of light reflected by the shaded area. The painter has to solve the shadow equation, calculating how much he must darken a certain zone of the painting so that the relationships among the various levels of brightness are enough to suggest natural chiaroscuro. It's very easy to go overboard. Novice painters tend to overdarken the areas representing shadow, making black puddles that have no relationship to the light. The representation of shadow must tell of the light that creates shadow; the slightest error freezes the shadow and makes it speak only of itself.

The difficulty in getting the proper dose of luminosity is compounded by a geometrical difficulty. Shadow also speaks of light by indicating where the light comes from. The physical lines of light in the picture must be cast so as to restore the spatial relationship among the light source, the object casting the shadow, and the cast shadow on a screen or on the ground. These projections are based on the same math that underpins perspective. An adequate solution to the problem of cast shadows requires, first, a solution to the problem of perspective.

Contrivances

Precision about light and perspective isn't everything; you can get pretty far without worrying too much about that. Or at least, that's how people got along for quite a while. The painters decorating the house of Augustus on the Palatine Hill in Rome showed how in Pliny's time, at the end of the first century B.C., shadows could be used to uphold the illusion of space even without the use of a theory of shadow projection. Since this was the emperor's house, we can reasonably assume that it is probably among the best examples of the art of the era. In the place called the Room of the Masks, the frescoes show architectural elements opening onto the outdoors: our gaze is accompanied through this space by a receding series of slim columns that frame a window beyond which the outdoors can be glimpsed. The images fool the eye: perhaps the artist was inspired by theatrical backdrops—perhaps he was a scenery painter (and the painted masks scattered throughout the fresco might be his signature). The perspective of the columns is reinforced by their shadows on the wall, which indicate that the colonnade continues, reaching into our space on this side of the painting. It's important to note that these are *cast* shadows, not simple chiaroscuro effects.

Almost perfect shadows: the Room of the Masks in Augustus's house on the Palatine Hill in Rome

Although this technical contrivance makes a remarkable three-dimensional effect, the shadows don't seem actually to have been constructed geometrically. A stereotyped formula, possibly copied from life, was applied to make the little columns' shadows—a formula chosen and then reproduced indiscriminately because it had served well in some other circumstance. But of course not all circumstances are equal. To start with, at one spot in this room the formula is repeated mechanically even though part of the paneling between the columns is missing, the part that would be required to cast that shadow. In another spot the shadows of a little railing make a curve that's optically impossible. We may think that there was a division of pictorial labor: that a colorist came in after the lines were sketched, and at the same time a shadower was brought in, who worked by copying carelessly from a sample book. And without worrying too much, either, about the coherence of the scene: sometimes the shadows all come from the same light source (a fact that helped the restorers a great deal when they had to reconstruct the fresco from fragments); but sometimes the illusion in one single area is more important, and the artist seems to forget that no light can create the scattered shadows he painted.

I use this example, and I dwell on these mistakes, to make a more general point. In the history of painting, shadows have been discovered and rediscovered by trial and error, and the dialogue with shadow can help us to capture the mind—in this case, the painter's mind—in midstream, in the hope that some slipup or forgetfulness will reveal a trail through the labyrinth of cognition.

But we have to proceed very cautiously. The cognitive sciences deal with individuals and with what happens in their brains, and the majority of the artists we study are no longer here to explain what they were really thinking. A generation of art historians and psychologists in the wake of Ernst Gombrich and Rudolf Arnheim have shown that the cognitive interpretation of a work of art is the result of a delicate balance of what we know about perceptual structures and what is passed on to us about the culture in which the works in question were created. In the best circumstances we can read the artist's cognitive process right in the work. The work done by an artist to solve a particular visual problem always risks leaving some kind of trace, some error that exposes his thinking.

Shadows That Fall Short

The history of Western painting abounds with systematic errors; in the context of this discussion one might call them shadows that fall short.

Look, for example, at the *Nativity* by Fra Filippo Lippi (1406–1469) in the Cathedral of Spoleto, which the artist painted at the end of his life (overleaf).

Beneath the three angels at left, a pole juts out from the wall of the ruined building that houses the manger. The pole casts a short shadow that seems to run down from left to right. The far-right wall (which makes an angle with the first wall) also has small protruding beams casting shadows that look to us as if they're dropping from left to right. Whether we examine the shadows one by one or look at them in the context of the whole painting, we would say that the light must come from a single spot at the top left. But if we try to pinpoint where the light source is, we realize that there is a subtle inconsistency. The direction of the shadows on the right wall requires a source on the *far* side of the house (because the shadows come *toward* us). On the other hand, the shadow on the left wall requires a source located on *this* side of the house (otherwise that shadow couldn't fall back against the wall).

In Fra Filippo Lippi's Nativity *the shadows cast by the poles are* inconsistent: *there's no* single *source of light that could cast them. Still, the scene doesn't look unrealistic: our perception accepts a lot of what reasoning declares to be incorrect.*

At a glance, it's not immediately clear that something's wrong. Perhaps the inconsistency is almost imperceptible because the shadows are so far apart. Or maybe it's because they are all pointing from upper left to lower right of the painting's surface. Or maybe, more simply, it's just because the inconsistency is forgivable. I purposely chose a term from the field of logic when I spoke of the *consistency* of the shadows. There is an intellectual—as well as a visual—aspect both to the act of drawing and, in some cases, to the interpretation of the work. Just like the frescoes in the house of Augustus, the painting in Spoleto shows that painters can render the effect of light in a space without worrying too much about overall coherence. Faced with a painting, the observer's mind needs only a few clues to reconstruct the scene being represented, and the reconstruction isn't hindered (too much) by any contradictions that may arise between the shadows. In order to discover the contradiction, one has to hypothesize: *If* the right-hand shadow goes this way, *then* the left-hand shadow cannot go in this other direction. This sort of reasoning shows that shadows are objects that make us think; that they speak not only to the eye, but also to the mind.

Surprising Shadows

Some narratives give shadows a leading role, so the painter cannot avoid representing them, even if it seems that he would be just as happy not to. At the beginning of the 1500s, Luca Signorelli painted some scenes from Dante's *Divina Commedia* in the Cathedral of Orvieto. In the fifth canto of the *Purgatorio* the souls of the dead are amazed to see that Dante is not one of them, because he casts a shadow.

Luca Signorelli, Dante's Shadow, *San Brizio Chapel, Orvieto. "When they became aware I gave no place / For passage of the sunshine through my body, / They changed their song into a long, hoarse 'Oh!' "* (Purgatorio, *V, 25–27). "And saw them watching with astonishment / But me, but me, and the light which was broken!"* (Purgatorio, *V, 8–9).*

The dead souls, being shadows themselves, do not cast shadows. Dante is an alien among the souls, and he is unmasked by his shadow. The scene is rendered dramatically: Dante turns around to look, almost with disgust, at the shadow that betrays him.

But the representation is puzzling.

In this fresco *the souls also cast shadows.* Their shadows are certainly different from Dante's. The few shadows that appeared in ancient Roman painting and mosaics virtually disappeared from art by the time of the Renaissance, leaving behind only a little "anchoring shadow," an abortive shadow stuck to the figure's feet. (This is not just a pictorial invention: when the sky is cloudy, you too cast a vague little shadow that reaches just a few inches beyond your feet.) This served only to fit the object into its space. When painting a person, one must indicate where the person touches the ground; otherwise it looks as if he is floating somewhere above the terrain.

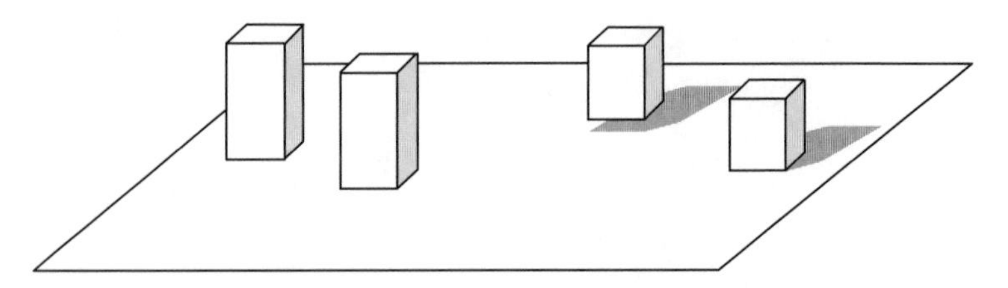

Without an anchoring shadow, the shapes don't have their feet on the ground: they seem to hover. A shadow indicates the presence of a surface and the object's distance from the surface. It also indicates something about the slope of the surface.

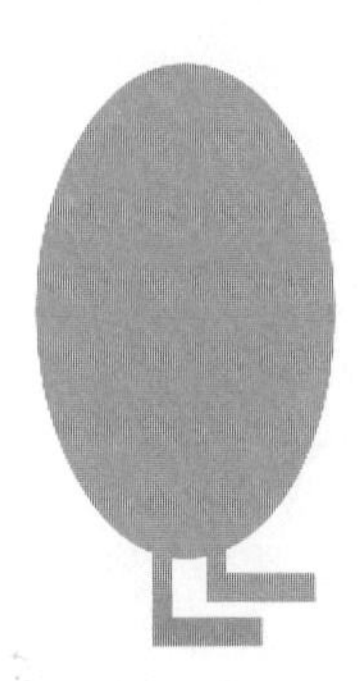

Dante's odd shape, as derived from the shadow painted by Signorelli.

The point where the shadow touches its body is also the point where the body touches the ground. Because it's so simple and economical, the anchoring shadow is an effective and universally useful pictorial tool. In painting the scene of Dante, Signorelli apparently had to choose whether to be faithful to the text (souls without shadows) or to make the picture legible (souls with their feet planted on the ground), and he chose the latter.

Looking closely, we notice that Dante's shadow is also unsatisfactory. It's a strange hybrid: Dante is right to look at it with such disgust. It looks like it's cobbled together from two anchoring shadows tied to the poet's shoes, plus a big oval shadow—completely abstract—that has no relation to the shape of Dante's body and cape.

Signorelli was working without a theory that would have shown him how to solve the problem of the shadow's shape; he just muddled through. He strapped the stereotypical anchoring shadows to an innovative shape that's almost an abstract figure, a hieroglyph.

Magic Shadows and Transparent Bodies

Attempts to overcome the obstacles posed by cast shadows have been sporadic, and in the absence of a good theory some artists ran into problems that were conceptual rather than pictorial. One example is Masaccio's fresco of St. Peter healing cripples with his shadow, in the Brancacci Chapel in Florence. The legend harks back to the strange powers of shadow. St. Peter is shown walking past three fairly rundown people: two of them have already been brushed by Peter's healing shadow, and they have risen to their feet, cured. The third person is caught at the moment when the shadow is starting to work on him.

In Masaccio's St. Peter Healing with His Shadow *(1425), the saint's shadow passes right through the miraculously healed man, without shading him.*

The shadow's magic is the theme of the painting, so Masaccio cannot fail to show the shadow. But he's not quite sure about how it should fall. The third cripple is in the shadow's trajectory, but the shadow—even more magical than we thought it was—seems to slide right under him without darkening him. The light and the shadow part ways, and the supplicant stays bright. Or maybe we're supposed to believe that the bodies of both the saint and the supplicant are magically transparent: the supplicant is transparent to shadow, and the saint casts a shadow even though he is transparent to light. This strange effect appears again on another wall of the chapel. Here the shadows are not part of the story, so they appear only as a special gift from Masaccio to the observer. (Masaccio probably had his own reasons for painting shadows cast by people but not by trees and houses, which would be just as likely to cast them.) Even though the shadows in this scene are not therapeutic, they behave just as magically. They are only two-dimensional traces on the ground; they slide right through people without shading them, and they are trodden on by people who cast shadows of their own but who don't ever shade their fellow humans. The dark rags that stick to the ground and slip humbly beneath objects are the most abstract distillation of shadow: black patches.

In Masaccio's Tribute Money, *shadows pass invisibly through bodies and then hit the ground.*

How to Make an Object out of a Shadow: Black Patches

Let's go back for a moment to Pliny and his legend of the origin of painting. To review the sequence in detail:

One: The shadow of the subject being painted (which stands in profile against the light) is cast on a wall. Nothing strange about that: the subject and the shadow get along fine—the latter is obviously a shadow, a projection.

Two: The portraitist traces the profile of the shadow with a heavy line drawn with charcoal or with a black brush.

Three: The subject moves away, and his shadow follows him away, while the profile stays on the wall.

This is more or less what we understand from the story. But a key element has slipped in, unobserved.

You have to *try doing* the sequence to notice a phenomenon that cannot be imagined, which escaped Pliny and many other art historians and commentators on this story (there is no record of anyone having mentioned it). The second phase is the crucial one; it has enormous visual impact. Tracing a line around a shadow *transforms the shadow.* All at once the shadow no longer belongs to the portrait sitter: *it becomes a colored patch on the wall* and takes on a life of its own.

I strongly encourage the reader to try it. Put an opaque object—for example, a little statue—on a sheet of white paper, and set it in the sun or under a bright light so that its whole shadow falls on the paper. Now take a thick black pen and trace the shadow's outline. Look at the result: *you no longer see a shadow,* but only a gray shape. If you move the statue slightly to the side, the shape disappears and the shadow reappears. What has changed? The area inside the line has *taken on a color of its own*—a dark gray. This gray is very different from the gray of the area before it was outlined. In fact, if someone had pointed to the shadow zone before the line was drawn and asked you what color it was, you would have said "white" (meaning the zone, not the shadow). Simply putting a piece of paper in shadow doesn't change the paper's color. Our visual system is able to correct for and to cancel out the effect of lighting. When shadows turn into patches, they completely stop being shadows and become something else.

This contrivance of tracing lines around shadows was much studied in the psychology of perception at the beginning of the 1900s. But painters considered it a mistake that should be avoided. Already, many centuries ago, Leonardo da Vinci observed that an outline ruins a shadow, making it look "wooden." Ewald Hering (1834–1918), one of the pioneers of the study of vision at the end of the 1800s, rediscovered Leonardo's effect in his laboratory. He explained it by saying that the line *erased the penumbra* and made the shadow lose its property as a phenomenon of light.

The odd thing is that even when the shadow/patch is moved away, the outline is enough to indicate that there was a shadow. In 1969, perceptual psychologist J. M. Kennedy studied ten thousand drawings from different artistic traditions and noticed that not one of them portrayed a

The statue's shadow on the paper doesn't make the paper gray; we still see it as a white paper with a shadow on it. But if we trace a line around the shadow area, it suddenly becomes a gray patch. (A photograph inevitably weakens the effect.)

cast shadow using only an outline. Is this a pictorial convention, or is it a problem of visual perception? Kennedy showed the subjects of an experiment a series of drawings with outlined shadows. The participants had probably never seen pictures of shadows reduced to simple outlines (at that time graphic designers hardly used images based on the high-contrast or solarized photography that reduces shapes to simple outlines). And yet the subjects had no trouble recognizing that the outlines were shadows. But when a pattern—such as a wallpaper print—filled an outline, the subjects got confused and no longer saw it as a shadow. Kennedy concluded that shadows have three distinctive characteristics: absence of internal detail, location on a surface, and two-dimensionality.

So if shadow gave birth to painting, it did so thanks to the line that separates it from light. And then the line went on to live a life of its own.

Cognition of Shadow and Patches

It's hard for painters to draw shadows, and they have to be careful not to make the "wooden" dark patches that Leonardo feared. Oddly, even though shadows are *not* gray patches, painters must learn to *see* them as gray patches because when they reproduce them on the canvas, shadows (like any other object) are ultimately represented by patches of color. Painters use a simple trick. If you look through a cardboard tube at a perfectly white teacup that's partly shadowed, you can see it as two-tone or patchy. The border between the lighted part and the shadowed part now looks like the border between two colored areas. It no longer looks like a shadow. Which means that it's not enough for something to be a shadow in order to look like a shadow. But since our eyes *always* register a gray patch, why don't we always see a gray patch instead of a shadow? Evidently, the brain "cancels out" the gray areas that correspond to the shadows in the picture. And it does a great deal to reconstruct the shadows corresponding to the canceled-out spots, and to put them back in place—as shadows.

Why does the brain use so many resources to dismantle and reconstruct the visual scene?

The fact that the process is instantaneous indicates that the whole thing is an atavistic holdover, something that has been with us for a long time, something that's part of our legacy of animal cognition. It helps the brain to be able to differentiate between what's transitory and what's permanent, in order to keep an eye on changes in one's surroundings. A white object with a gray patch is a two-tone object. A white object with a shadow falling across it is a monochromatic object and must be seen as such. A patch is a (semi)permanent characteristic of an object. But a shadow is a completely transitory characteristic. Is something happening to the terrain I'm walking on, or is it just a passing change caused by my shadow? When one can automatically differentiate between the two circumstances, one needn't waste precious time worrying about it at every step.

When the complete visual image is generated—the film of reality that we see—the brain fills in the area from which the gray patch was erased with *three* pieces of information: the color (without shadow) of the object in the shaded area; the presence of the shadow; and a reassuring message: "This is a transitory object: don't waste time examining it."

The Contents of the Lost Book

In any case, painters must study shadows if they want to represent them convincingly.

The difficulty of representing shadows runs through the centuries; Renaissance treatises on painting show evidence of a real hand-to-hand struggle with shadows. Leonardo was one of the great generals in this battle. He seemed to have everything he needed to make shadows in paintings, but he couldn't put the pieces together into an organized theory. Here's the story. Leonardo inherited a system in which both shadow and chiaroscuro (and, to some degree, cast shadows as well) were canonized in a rule of three: *use three gradations if you want to obtain all the effects of light—pale for the illuminated part, intermediate for the transition, and dark for the shadowed part.* This very simple rule satisfies the eye without making the painter work too hard; with a minimum of graphic tools you get a maximum return of visual effects. Around 1490, Leonardo decided to go to the heart of the problem and figure out how shadows really work. The resulting theory set shadows in opposition to light, but with the two entities referring back to one another. The strangest part of Leonardo's theory is this: that shadow itself is active—it emits rays like light, called *shadowy rays.* Let's allow him to speak for himself:

> Shadow is the lack of light. Given that shadows seem to me to be indispensable in perspective, and given that without them it's hard to understand dark and opaque bodies, and also what's within the shadow, and its borders . . . for all these reasons, I propose and enunciate in my first proposition about shadow that every human body is encircled and clothed all over with shadows and light, and I'll build my first book around this. Furthermore, shadows are different qualities and degrees of darkness, because they are abandoned by different quantities of luminous rays. We call these original shadows, because they are the first shadows to wrap around the bodies that they're attached to, and I'll construct my second book on this. Shadowy rays come from these original shadows, and they spread through the air and they are as varied as the original shadows that they derive from; that's why I call these derivative shadows, because they are born of other shadows; and I'll make my third book about this. These derivative shadows have different effects according to where they strike; and here I'll make the fourth book. And because the striking point of the derivative shadow is always circled by the strikes of luminous rays, and—reflecting and returning to its source—it meets up with the original shadow, it mixes with it and turns into it, changing the other's nature; I'll build the fifth book on this. Furthermore, I will make a sixth book in which we will investigate the various modifications of the resulting reflected rays.

Only a draft remains of Leonardo's "Book of Shadows." It's not clear whether Leonardo ever progressed to a more final version of these chapters; we are left with only some traces of his theoretical studies of shadow, but they are not enough to indicate a whole, organized work. The project suggested a natural history of shadow from the moment that light hits a body, through the shadow's projection onto a screen, and beyond. We can follow Leonardo's history with a diagram that looks innocent enough but which hides a bizarre concept.

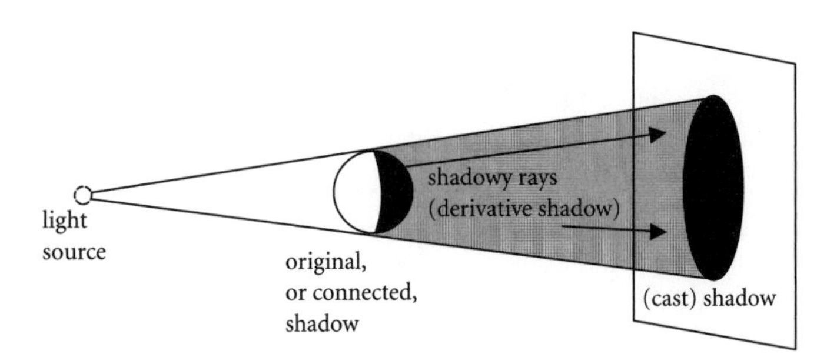

The original shadow (called a *connected* or *attached* shadow) is the one that sticks to the body. It's not a passive phenomenon insofar as it deflects light rays and emits shadowy rays (*derivative* shadows), which in turn strike a screen and create a projected shadow (or *separated* shadow: in the pictorial language of this period, such a shadow was called the *sbattimento* or the "hit"). Shadowy rays are not visible by themselves. Leonardo says that if we illuminate a sphere from below, we make a bundle of shadowy rays that's invisible and that gets lost in the sky—in just the same way that we cannot see the shadows thrown into the night sky by the moon and by the earth.

Shadowy Rays: Shadow Is a Black Light

This business of the shadowy rays is curious. Naturally, the theory is false; even Leonardo knew it, for he did say that shadow is a lack of light. But a false theory can be very persuasive. Indeed, the pseudo-concept of shadowy rays explains a *geometric* property of shadow by attributing to it a particular *physical nature.* This is a roundabout way of getting to geometry via physics. The ambiguity lies in the concept of projection. Light spreads physically by making a geometric projection; shadow is *only* a geometric projection—it's not propagation. But the confusion isn't completely unjustified, as we can see if we consider *colored lights*. If we could say that shadow is a colored light, we could beautifully unify two optical phenomena. The concepts of shadow and colored light leach into one another, seducing the imagination and allowing us to consider shadow as just a borderline-case lighting situation. Such an interpretation reconstructs one of Leonardo's possible arguments. And there are some text passages that support this. In the *Codex Atlanticus*, for example, folio 658 recto shows Leonardo's studies of superimposed shadows and, side by side, superimposed colored lights. It has been observed that these studies of Leonardo's are not the result of empirical observation but of a priori reasoning: Leonardo was much less of an experimenter than he is usually thought to have been. In this case, Leonardo grasped one aspect of the naïve conception of shadows. At this point I'm compelled to offer another shadow brain-teaser.

Brain-Teaser Number Nine: The Green Ray

I'm holding a disk of green glass (say, a transparent green ashtray) about five inches above the tabletop, below a lamp; on the table is a green patch.

Is the glass casting a *cone of green shadow* or a *cone of green light?*

Let's say it's a green shadow. But there's one problem: if I lift the ashtray up so it shields the lamp completely, the room will fill with green light, and in that case I wouldn't say that there was green shadow in the room. So let's start again and say that the green patch on the table is a patch of green *light*. Fine. Now let's pick up a gray glass ashtray. Would we say that this makes a cone of gray light? And if the glass were even grayer, would it make dark gray light? And if the glass were opaque, would it make black light? Wouldn't that just be called shadow? *So what's the difference between light and shadow?* Maybe Leonardo was right after all.

This is the dilemma: If we say that the green glass casts a green *shadow*, we must conclude that a lamp screened with glass makes a green shadow filling the whole room—although we would rather say it makes a green light. But if on the other hand we say that the green glass casts a green *light*, we have to conclude that gray glass makes a gray light (getting darker if the glass gets darker), and not a gray shadow. None of the descriptions we began with works very well. Leonardo stumbled over the ambiguity of the concept of shadow when he wrote about shadowy rays.

There is one solution to the dilemma that takes into account the laws of physics and goes beyond commonsense concepts. The green glass filters out all wavelengths except those corresponding to green. So it shadows the table from all components of light that are not green. The gray glass filters some of the light in all wavelengths. Because it does not choose a particular type of light, we can say that it makes a partial shadow compared to white light. So both the green and the gray glass make shadows, and we were wrong to say that the green glass produces green light. But then what happens when the green glass completely screens the lamp? We have to say that it produces shadow, and not that it produces green light. I surmise that we don't really want to say that it makes a shadow because the ordinary concept of shadow calls for a shadow line, a demarcation setting it apart from the light, which is not available in the case of the glass completely screening the lamp.

There is an experiment that can indirectly confirm the fact that shadows and colored light are false friends. Imagine a column in the middle of a theater stage. At the left of the stage a spotlight bathes the column in red light. At the right a spotlight bathes the column in blue light. The column casts two shadows in opposite directions. I have sketched the scene below. Now take blue and red pencils and fill in the correct colors for the shadows.

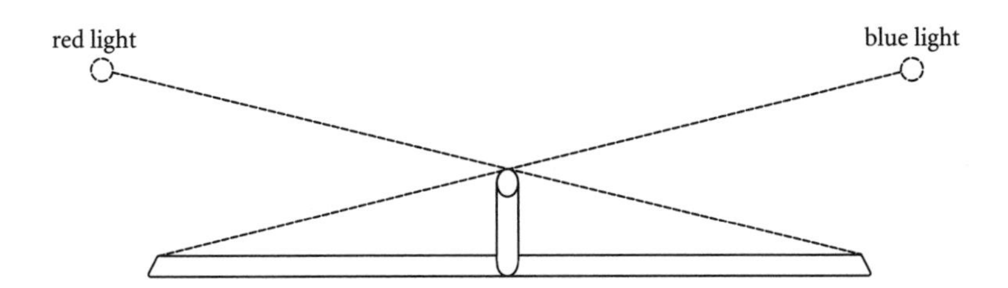

If you colored the left-hand shadow blue and the right-hand shadow red, it means that you are treating shadow like colored light. In reality, the left-hand shadow would be red. Indeed, it's the only stretch of the whole stage that doesn't get any blue light; it's lighted only by the red spotlight! (Likewise, the right-hand shadow is blue.)

There's one more cognitive oddity. Why do we say that shadows are *cast?* Is it because we imagine a *movement* from the object to the shadow? Perhaps we use light as a model for shadow? No. The similarity with light is useless: between a light source and a pool of light there is a transfer of energy that travels in one direction. But *between the object casting a shadow and the shadowed surface there is nothing moving,* so there is no preferred direction (unless we take seriously the idea that there's such a thing as "the speed of shadow"!). So why does the mind invent a directionality, and refer to *projection?* The linguist Len Talmy has studied this and other images of movement hiding within our linguistic descriptions. His theory is that we tend to think of a direction from the shadowing body to the shadowed surface because this suggests that shadow is an object dependent on the body. The asymmetry of the movement serves to describe an asymmetry of dependence. I think this theory sounds right. And I'd like to add a charming detail: denying that shadows are really projected physically, Talmy jokes that shadow projections are not a physical fact because there is no shadow-thing that corresponds to photons—no "shadowons" around. (Another possible name for them would be "skions.") Oddly, Talmy reinvented Leonardo's pseudo-concept of shadowy rays, in modern terms.

What the Lamp Sees (What the Lamp Doesn't See)

Leonardo was brimming with ideas, but he was not keen on developing a mathematical theory of shadow projections; instead, he came up with an inadequate physical theory. That's unfortunate, because he was very clear that shadow projections and perspective projections are two aspects of the same thing. Indeed, he wrote, "in matters perspectival, a light source is no different from the eye. The reason why light is no different from the eye in losing sight of a thing behind the first object is this: you know that the visual ray is similar to the shadowy ray in terms of speed and converging lines." And he drew a picture that couldn't possibly be clearer:

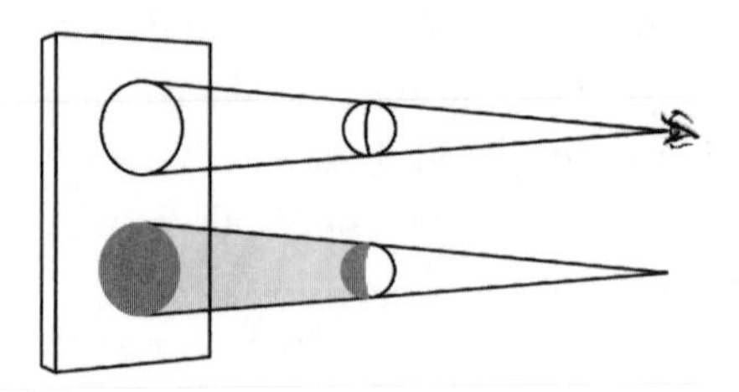

Copied from Leonardo: the eye is like a light source.

This is a strong analogy. *There is shadow where the light cannot see.* Indeed, shadow *is* exactly what the light source *cannot* see. A lamp sees only the things that it illuminates; objects in shadow are behind lighted objects that screen the shadow from the lamp's perspective. For this same reason, when we examine a shadow, we discover the profile of things from the point of view of the lamp. I have a suggestion for entertaining yourself if you ever happen to be sitting through a boring conference. When you read the words cast by the overhead projector, you can imagine that you're standing in place of the projector bulb; when the presenter moves his transparent slides, you'll see his hands as if you were right there. Shadow allows us to see a point of view other than our own without even getting up from our seats. When an object and its shadow are visible at the same time, it's as if we're getting two different points of view. It's *almost* like seeing with two separate eyes at once— "almost" because the shadow and the object are not synthesized into one single image, as is the case with real binocular perception. Here the synthesis is the result of reasoning. We might say that it's binocular vision for the mind's eye.

At this point it's only a short leap to link intellectual comprehension of shadows to the actual sketching of a shadow: we show a person a light source (unlit) and an object, and we ask him to sketch the outline of the shadow that would appear if the light were switched on. It's not easy to do. The psychologists Piaget, Inhelder, and Ascoli did experiments to examine how children develop the ability to conceive of a shadow's shape based on the positions of the light source and the object. In the earliest phase the drawings showed the shape of the object as seen by the child, not as seen by the lamp; only later did the children take into consideration the lamp's point of view.

This drawing exercise reveals the profound affinity between perspective and the projection of shadow. It's a very close relationship. Since they are projection-objects par excellence, shadows were unlikely to have been ignored by the masters of perspective.

XVI *Shadow Webs*

It's hard to judge a shadow.
—Cyrano de Bergerac

According to the myth that Pliny told, shadow gave birth to an illustrious child—painting. During the Renaissance, shadows seemed determined to confirm this story, and indeed they might go even further, taking credit for the invention of pictorial perspective.

To make a picture in perspective—for example, a picture of a line of people walking toward us—you have to make some of the people look as if they're closer and others farther away. The simplest way to achieve this is to draw the "farther" persons smaller than the "closer" ones. The risk is that this shrinking (or foreshortening) can upset the perspective. Once you begin foreshortening some things, you have to go on foreshortening everything in the scene, and keep gauging the effect of the foreshortening of each object. One little error and the effect is ruined.

So it would be helpful to have a method that would resolve the problem of foreshortening once and for all, a *construction* for every object in the painting. The most successful constructions try to reproduce, within the picture, the way objects *line up* in real life, or to give the impression of plausible alignments. (In the two drawings on the next page, the gray figure seems like a child or an adult respectively, because his head doesn't line up with the other heads in the group.) Pre-Renaissance paintings often give the impression that *some* alignments were respected, but not all of them. Renaissance perspective does succeed in maintaining all the alignments, or at least suggesting to the viewer that all the alignments have been maintained. The secret of this success is that Renaissance perspective works as a *projection* from reality through the painting and toward the eye of the viewer. In the language of mathematics, which describes various types of projection, this kind is called a central projection. (There is one "center," a spot from which the projection rays emanate.) The central projection both foreshortens objects *and* maintains alignments.

A child in front of a line of adults and an adult behind a line of children. Actually, the gray figures in the two lines are the same size.

Well, shadows from a lamp also result from central projection. It would be *possible* to arrive at the laws of perspective by studying shadows. Is this just a theoretical possibility—or did perspective enter painting because of the careful observation of shadows?

Nothing is very clear about the sequence of events that led to the Renaissance-era discovery of the methods that permit perspective construction. Theories abound. For example, many people have pointed to windows or mirrors as tools for discovering the geometry needed to construct the image. It's a simple idea: carefully study what appears in the window or the mirror and you will learn how to construct good perspective images.

The art historian George Bauer, however, champions shadows. His too is a simple idea. He says that painters used models for the scenes and the people they intended to paint; they would project the models' shadows on a screen and scrutinize the intersections of their alignments. By studying shadows, which are projections, the artists understood the principles of perspective.

I'll try to evaluate the two theories. My comparison is purely conjectural. I will consider what windows and shadows *could* do, not what people think they actually did.

Naturally, it's one thing to take advantage of shadows in order to understand the laws of perspective; it's another thing entirely to use shadow-casting models in order to make your drafting easier when you're in the process of making a painting. We have to keep this distinction in mind.

The Window and the Mirror

Before examining the historical proofs in favor of shadow, let's look at the logic of Bauer's proposal. After all, one might say, what could be simpler than discovering the laws of perspective? Just slip a piece of glass or a mirror between us and the scene, trace the outlines of the objects as they appear through the glass, and then examine the properties of the figures that you've traced. You can discover foreshortening and the necessary alignments.

People often speak of paintings as windows on the world. Albrecht Dürer (1471–1528), in a famous drawing, went so far as to furnish instructions for the pictorial use of windows.

You need a painter and an assistant. Imagine the two of them at work. The painter has driven a nail into the wall behind him and he is staring at a certain spot. In front of him is a latticed framework sitting on the

Dürer's cumbersome mechanism for practicing perspective, from Underweysung der Messung, *Nuremburg, 1525*

worktable, with a window that opens and closes; the drawing paper is attached to it. The assistant holds a string running from the nail to a point on the object that is to be drawn, a lute located on the far side of the window. The painter, with a firm hand, marks the spot where the string goes through the framework. Then the window is closed and the point fixed by the painter is imprinted on the paper. After a certain amount of back-and-forthing by the assistant, a constellation of points appears on the sheet and coalesces into the shape of a lute. Quite a lot of bother.

It's important to understand why this analogy is misleading.

With the window method, the projection surface *stands between the painter and the scene.* The painting opens onto the scene just like the window does. The analogy grabs the imagination, but the similarity is very superficial. Windows and mirrors are quite different from paintings *because of visual parallax.* If you move to the right or the left in front of a window or a mirror, you change your relation to the various elements in the scene you see. The flower seemed to be on the right of the sword; now it looks as if it's on the left. If you move to one side while looking at a painting, your movement doesn't change these relationships. (At most it will change the conditions for good visibility—you can't stand too far off to the side.) And indeed Dürer's painter must have a steady hand and an unwavering gaze.

On this score the method of generating images with shadows has a certain advantage. The projection surface *is on the far side of the scene.* The painter doesn't have to stay put in his visual posting: immobility is only the lamp's job. The painter can get up close to observe the web of the shadows' intersections; he can trace the lines of construction and examine them while moving around freely. And he can do so comfortably: because the shadows aren't bound to his point of view, they don't suffer the effects of parallax as they would in the window or the mirror. The painter reproduces not his own point of view but that of the lamp. And he doesn't need his poor assistant; or rather, the shadow is his assistant.

But How Did the Discovery Happen?

Like a detective creating a model of the crime scene, I managed to set up a shadow scene using a candle. The candle was on the floor, three feet away from a white wall. Between the candle and the wall I built a space flanked by some piles of books (buildings). A dozen pencils balanced precariously on end were lined up along the two sides of the "road" (as lampposts). I lit the candle, and on the wall appeared a simple but convincing perspective image of a city avenue.

It's worth trying to reconstruct this scene at home. You get a marvelous effect if you use a flat-bottomed candleholder that you can slide across the floor: the dynamic spectacle it casts on the wall looks like the animation found in a road-driving video game. The movements of the candle correspond to impressive changes in perspective on the wall. It's as if you're inside the scene.

Later, I tried bridging the street by placing a bicycle basket across two of the book piles on opposite sides of the street. The basket was made of widely spaced wires, and on the wall *the image of a coffered ceiling* was cast at once.

One could imagine that a process like this was discovered by chance; after all, painters' studios were likely crowded with all kinds of equipment, and a light source probably happened to be in the right place occasionally. Some image cast on the wall must have looked particularly pretty or inviting. The outline of the shadow got traced, and the foreshortening looked interesting. If someone used a network of squares like my basket, he would have seen the geometric model for the countless regular floor patterns and ceilings that adorn the paintings of the Renaissance. And he would have a model that was full of information. With the design on the wall he could have examined all the properties of this grating, and he would have understood how to build it. Once you've figured out the method, there's no longer any need to repeat the shadow stratagem; the construction is simple and general.

This theory concerns only the *discovery* of the laws of perspective. There's no saying that shadow perspective is easier to use in *practice.* If the painter cares only about representing a real scene, then reproducing it in a scale model—in order to then project its shadow onto a wall or a canvas—certainly isn't the easiest or cheapest way to proceed. From this standpoint a perspective window seems more advantageous, if for no other reason than because it's portable and you can set it right in front of the scene you wish to paint. But the role of Dürer's window is doubtful, if it even existed—and I would also add that it's just as clumsy as the shadow method. It's not at all easy to use. The assistant who chooses the salient points of the model-object must hold a string to every single spot. Any string more than a few yards long will start to droop, so it stops acting like a straight ray of light. It's impossible to send the assistant climbing to the rooftops. It's hard for the painter to keep the point of view. And, finally, too many points in the scene need to be marked. The method's direct and mechanical nature is completely overshadowed by its difficulties. Indeed, Dürer meant to impress his reader only with the mechanical and universal aspect of the latticed window method. The engraving that illustrates the method doesn't give instructions in the use of perspective. It serves only to explain the rules that allow the construction of a perspective view; it serves to legitimize these rules by drawing an analogy between lines of sight or visual rays and the strings that link the painter's eye to the objects. If you don't understand visual rays, Dürer seems to say, at least you can understand a string pulled taut.

Shadows and the perspective window are equal in both theory and practice. But cast shadows have some advantage over their competitor if what matters is to comprehend perspective. The painter can work directly on the image. The shadows are right on the wall and they can be examined comfortably.

Painting Is the Shadow of Sculpture

So far, so good. But the fact that shadow experimentation *could* explain the origin of a practice doesn't mean that the practice *actually was* born from it. We have to look into the scanty surviving documentation. George Bauer claims to have detected a trail leading to Leon Battista Alberti (1404–1472), author of the treatise *On Painting* (from 1435; one of the seminal texts of perspective, which inspired Leonardo). In his book *On Sculpture,* Alberti recommended that a painter practice drawing parts of the body and notes that "the profile bordering the surface that we see gives a determined point of view and separates it from what we do not see; if it were traced properly on a wall, it would produce a figure exactly like that made by a shadow when it intercepts the light, if the light source were located in the same point in space in which the observer's eye was located previously." It's not a question of imagining the shadow that a certain body might cast; we're talking about examining the perspective appearance of a body by examining its shadow. In Alberti's words, *painting is the shadow of sculpture.*

Alberti recognizes that shadows and perspective vision are both examples of the same type of projection that uses three elements: light (the eye), object, and shadow (the image). This idea had already occurred to the philosopher and scientist Biagio Pelacani, who straddled the fourteenth and fifteenth centuries.

Bauer believes that all theories about the origin of perspective are simply circumstantial, but that in the end many elements favored shadows. He says there is a "coherent argument: first, that the invention of perspective was preceded by some kind of experimentation; second, that experimental and practical applications of shadows were common and familiar at the time; third, that Pelacani before Brunelleschi's demonstration of perspective and Alberti after it both knew and said that a cast shadow was a perspective image; and finally, that the possibility of obtaining correct perspective images from cast shadows first suggested by Pelacani was recognized by Alberti."

The Shadow Club

At this point the Shadow Club has enrolled two new members, Alberti and Pelacani. But we should add some more. The art historian Samuel Edgerton has tried to demonstrate the importance of a minor adventure in the history of the Renaissance: papal secretary Jacopo d'Angiolo's voyage to Constantinople with Manuel Chrysoloras, a Byzantine scholar in his service. They roamed for years before returning to Florence in 1400, even suffering a shipwreck off the coast of Naples. But they brought their Florentine friends the bibliographic treasures that they had been sent off to gather. The travelers carried back chests of Greek manuscripts of classic texts to satisfy the cultural curiosity (as well as the collecting urge) of the humanists in their group at home. The most highly prized items were the eight volumes, twenty-six partial maps, and the great world map of Ptolemy's *Geography.* This work was known in Arab and Byzantine lands but unknown in Europe. The copy brought to Florence is important because this world map was the basis for the map created by the humanist Paolo dal Pozzo Toscanelli (1397–1482), which, after many vicissitudes, was used by Christopher Columbus at the end of the century to convince sponsors to finance his journey westward in search of a route to the Indies. (The map was wrong: it shortened the distance of the voyage by about 6,000 miles; but it also didn't include America, which Columbus stumbled onto in the area where he expected to find the Indies. Which proves that, strangely, sometimes two wrongs are better than one.) But Ptolemy's *Geography* has more than anecdotal importance. Ptolemy's map registered geographic facts and then went a step further, trying to frame them in a geometrically constructed space, and then to present this space as a picture. This last problem is tricky; it has been a trial for geographers ever since they began drawing meridians and parallels. The earth is (practically) spherical, but a spherical map is not easy to use; for one thing, it's hard to fold. A flat map is far preferable. But the projection of a sphere onto a flat surface inevitably deforms the shape of the geographic elements (for the same reason that a spherical map is unfoldable). Ptolemy suggested three methods for projecting the global sphere onto a surface, methods that furnish a grid for drawing unknown lands. Edgerton is particularly interested in one of these methods; he presents it as a foreshadowing of the perspective constructions used in the Renaissance. What was the next step? Toscanelli, also a cartographer, was an intimate of Filippo Brunelleschi (1377–1446), who was considered by many to be the geometric mastermind behind the story of perspective, and Toscanelli contributed to Brunelleschi's understanding of geometry. Giorgio Vasari writes in his *Lives* that "it happened that M. Paolo dal Pozzo Toscanelli, having returned from his studies and finding himself with friends at supper in a garden, invited Filippo along to honor them; hearing the discourse falling on mathematical subjects, [Filippo] formed a friendship with him and learned geometry from him. And although Filippo was not learned, he understood so well from his own practical experience that his reasoning frequently confounded Toscanelli." Even Alberti collaborated with Brunelleschi and Toscanelli—Alberti was described by Giorgio Vasari as a "very good mathematician and geometer"; he wrote about painting and sculpture and also about geometry and architecture (he did a new version of Vitruvius's books, which discuss sundials and geography, among other things).

Although this is still circumstantial evidence rather than proven facts, it's starting to look like there's a hidden path linking Ptolemy's projections to Brunelleschi's perspective. It also looks as if shadows played an important role in this connection.

More Shadowy than a Shadow

Toscanelli is a murky figure, so much so that the historian of philosophy Eugenio Garin called him "more shadowy than a shadow." Little is known about him, which is appropriate for a great "director" who worked behind the scenes; the little that is known about him makes him look like another Mr. Second-Rate, similar to Eratosthenes. The map that came into Columbus's hands was just one of his muddled scientific contributions. In 1456 he observed and described a comet that later turned out to be Halley's Comet. Around 1467 he took advantage of the building of the dome of Florence's Duomo to affix a pierced bronze plaque to the dome's lantern, more than 90 meters up, to make the world's biggest sundial. (On the cathedral's floor you can see a marble circle that marks the position of the sun at the summer solstice. The sundial's lines are now used to check the building's stability, because the cupola is always settling slightly. The delicate geometric paths of the planets are sturdier than any human construction.)

A completely different trail has been suggested to trace the link between Ptolemy's geography and Brunelleschi; this one too goes through Toscanelli. One of the texts familiar to the Florentines' Western, Latinate milieu was Ptolemy's *Planisphere.* According to the man of letters Alessandro Parronchi, Toscanelli brought from Padua to Florence a copy of a book titled *Questions of Perspective,* written by Pelacani (d. 1416), who was known as Biagio da Parma or "Doctor Diabolicus." This book is interesting because it explicitly links optics, perspective, and the projection of shadows, shifting the study of shadows from astronomy to optics. In his book Pelacani describes a way to depict a sphere by using the shadow it casts. This projection is inspired by the projection (called "stereographic") described in Ptolemy's *Planisphere,* which also serves as the basis for building an astrolabe. (This method is different from those found in Ptolemy's *Geography,* discussed above.) Pelacani refers explicitly to a treatise of Ptolemy's "on the projection of bodies on a plane," saying that it contains "a theory that comes to him through his knowledge of shadows." And he writes: "If you wish to obtain the illusion of a sphere in a room you must hang it in the middle of the room with a string that spins around an equinoctial circle. If at this point a lighted candle is brought close to the South Pole you will see, on the opposite wall, the candle, the sphere, and its flattened 'construction.' " The idea is the same as Alberti's and Leonardo's: put your eye in place of the light, and the image in place of the projection plane, and you'll get the perspective and the shadow, respectively.

This step is fairly important in trying to demonstrate that the Renaissance discovery of perspective derived from an understanding of geographic-style projections, and that shadows were the vehicle for this discovery. So I tried to replicate Pelacani's experiment. At first the results were unsatisfactory. For one thing, if you put a lighted candle near the south pole of a sphere hung the way Pelacani says, you don't at all see an image on the wall of the light source. Furthermore, the shadow of the sphere obviously makes a big black circle that takes up almost the whole ceiling, and it's hardly interesting for a painter—it certainly doesn't look like a sphere. Going back to his text, however, we can notice that Pelacani speaks of the flattened "construction" of the sphere. And he uses terms (like "equinoctial circle" and "south pole") that point not to an opaque ball but to an armillary sphere—that kind of celestial map with rings (*armillae,* or bracelets) representing the traces in the sky of the equator, the tropics, and the ecliptics. So I built a little armillary sphere of paper, and I brought the light close to the south pole; and indeed it projected some good construction lines on the wall—flat shadows of cosmic circles.

Fine. But what does all this show?

Not much. This projection is not the one Ptolemy described, even though it is mathematically equivalent to it. It's true that in both cases a sphere is projected by locating the projection point at the south pole. But in his experiment Pelacani projected onto a wall that was beyond the sphere's north pole. The stereographic projection that Ptolemy described in his *Planisphere* treatise works differently. The projection plane is located at the sphere's equator. But that means that only half the sphere—the part between the light and the plane of the equator—is projected as a shadow; the other half, located beyond the equator plane, is projected "backward" toward the equatorial plane. Ptolemy's projection is of the "shadow" type for the objects between the south pole (the

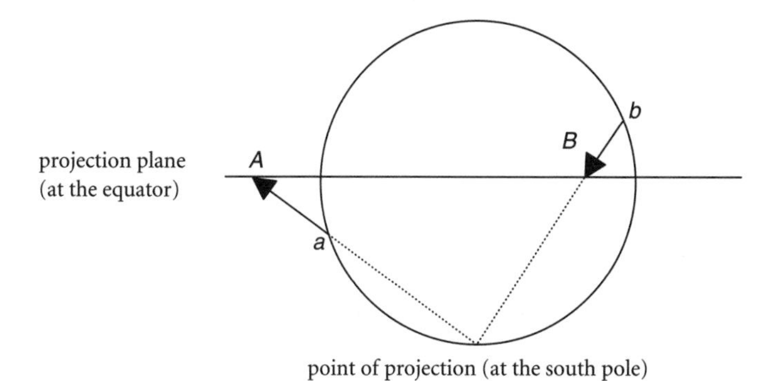

The stereographic projection from the Planisphere: *the sphere's point* b *is projected "backward" onto the equator toward its image,* B, *and toward the pole; but point* a *is projected "forward" onto the equator toward its shadow,* A.

projection center) and the equatorial plane (the point of projection is a candle), but it is of the "picture" type for the objects between the equator and the north pole (the point of projection is the eye of the observer).

This means that Pelacani's experiment could not be used for convincing Toscanelli (and by extension Brunelleschi) of the usefulness of the Ptolemaic method for constructing images. In short, even though shadows, projections, and images were explicitly associated in several texts known to the first great thinkers of the Renaissance, it's still not clear that the study of Ptolemaic methods of projection opened the door to perspective for these men. (After all, astrolabes, which use stereographic projection, had long been known; people were already practicing stereographic projection, so there was no need to wait around for Pelacani.) If Toscanelli and Brunelleschi learned anything from Pelacani's sphere shadows, they learned it independently of Ptolemy's geometry.

Let's sum up: If shadows helped in the rediscovery of perspective, it's plausible that they did so not because of complicated theoretical considerations about projection methods, but simply by offering themselves as examples of perspective images. The practice of shadows is easier than their theory. Still, the fulcrum for all these stories is the fact that both images and shadows are projections. This is interesting when we consider the mathematical structure of shadows, even apart from the historical hypotheses about their contribution to the development of painting. And for this we have to go looking for another missing book.

XVII *Shadow Lessons*

How can the shadow be straight when
The stylus is crooked?
—Abu al-Faraj

Around 1675 the philosopher Gottfried Wilhelm Leibniz (1646–1716) was hot on the trail of a book with a beautiful and mysterious title: *Shadow Lessons (Leçons de ténèbres).* We know only that it was published around 1641, and it included a contribution to geometry and perspective from Girard Desargues (1591–1661), a French mathematician and engineer. The title is a play on words referring to a Catholic Holy Week ritual that was popular at the time (night after night, a lecture and meditation ended in shadow after a candle was extinguished). The real subject of the book, however, was the relationship between perspective and the *science* of shadow projections. At last, we can use the word "science." A synthesis of perspective and shadow theory had appeared a few years before, in 1636, when Desargues published the *Treatise on Perspective Section.* This included an elegant theorem now known as Desargues's Theorem. To bring our discussion full circle, I should point out that Desargues also wrote a treatise on sundials.

Lining Up Shadows: Desargues's Theorem

Desargues's Theorem crosses the trajectory of mathematical tradition with the wisdom of the Renaissance masters of perspective; these two kinds of knowledge intersect on the tip of a sundial's stylus, then diverge and finally land, each of them frozen in their own lofty academic disciplines, such as descriptive geometry, perspective drafting, and projective geometry. Desargues made a *theory* out of perspective and shadow drawing, in that he enunciated a theorem that lies at the heart of every perspective *construction.* Desargues's Theorem is simple and surprising. Take a triangle and its shadow and extend the triangle's sides until they meet the extensions of the sides of the shadow; the intersections of the extensions all lie along the same line. If you draw an example of this theorem, or if you try to prove it with a real shadow, it always feels like a miracle when it works. Actually, Desargues's Theorem is among the few interesting mathematical propositions that ancient geometers could have demonstrated, had they only noticed it!

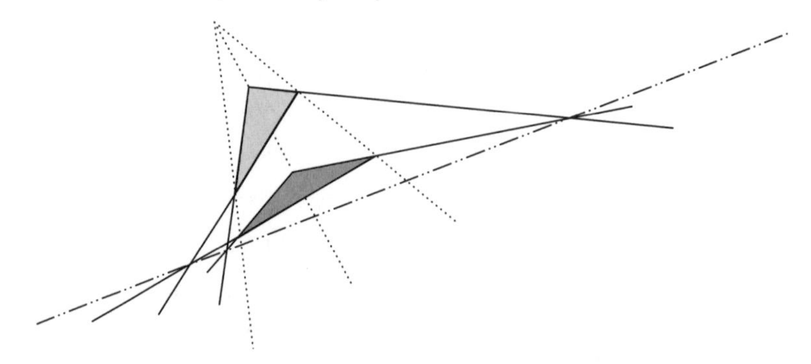

Desargues's Theorem. The points where the extensions of a triangle's sides meet the extensions of its shadow's sides lie along a straight line.

One interesting consequence of the theorem is that when one side of the triangle sits on the ground, all of its shadows are equally "good": all the intersections among the extensions fall along the side where the triangle comes into contact with its shadow, and the bonus of this constraint is that all the other points around it are "freed up": each of them is authorized to be the light source, the projection center.

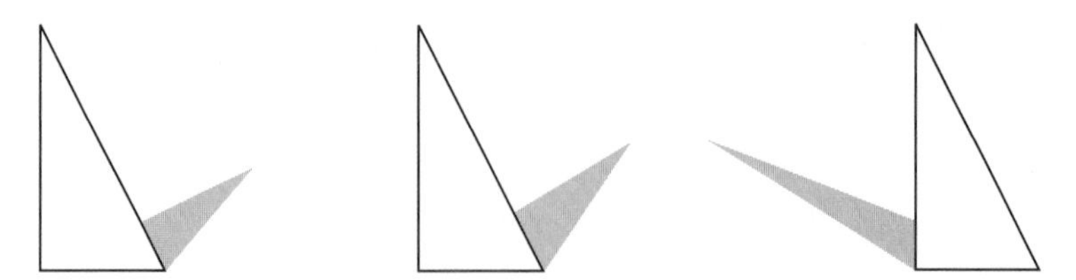

An easy drawing exercise. It's impossible to draw a wrong shadow if one of the triangle's sides is on the projection plane.

Naturally, the fact that a shadow is possible doesn't guarantee that it's also the appropriate shadow for a given light source. If the shadow is possible, it means only that its construction doesn't violate the principles of projection. But if the shadow is impossible, then it won't be appropriate for any light source, so you might as well start by constructing possible shadows. Here's an example of an impossible shadow:

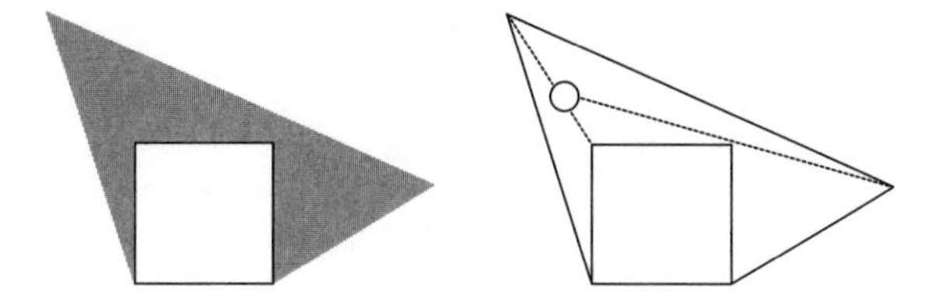

If you join the two upper corners of the square with their corresponding shadows, the intersection of the lines—the point where the light source should be—ends up *between* the square and the shadow, so it cannot project one part of the shadow.

Desargues's Theorem explains why one can *construct shadows geometrically.* In practice, knowing that there are constraints to the way the figures can be aligned actually makes the artist's job easier, because it creates reference points for the construction of the picture. The theorem can be applied to perspective pictures of objects in just the same way.

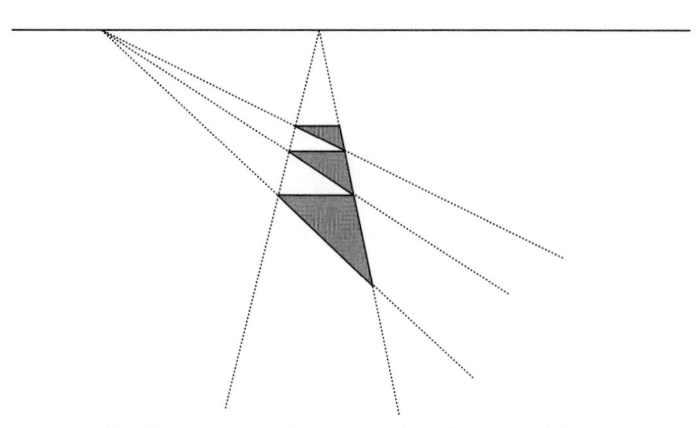

Renaissance perspective constructions show a special use of Desargues's Theorem; for example, you can suggest a tiled floor in perspective by simply lining up a series of triangles.

Drawing shadows becomes an intellectual activity. The artist relies on geometry to *calculate* the shadows.

Shadows Are Perspectives

Desargues's Theorem works equally well for shadows and for perspective paintings because shadows *are* perspective images. If we go back to Leonardo's draft, we'll recall that he suggested an affinity between shadows and images.

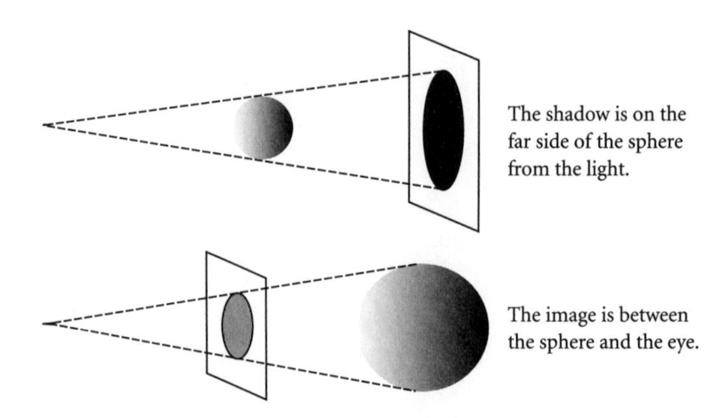

If we swap the eye for the light, the light ray for the visual ray, and the position of the sphere for the position of the section of the perspective cone, then we have *two examples of the same phenomenon.* Pictorial perspective and the projection of shadows are, in theory, *the same thing.* The only difference lies in the position of the projection plane.

Leibniz couldn't find the text of Desargues's *Shadow Lessons,* but he expressed the idea perfectly when he wrote, "The doctrine of shadows is nothing more than a reversed perspective, and is a direct result of this, if one puts the light source in place of the eye, the solid body in place of the object, and the shadow in place of the projection. The whole theory of sundials is nothing more than a corollary to a composition of Astronomy and Perspective."

If we delve into *shadow theory* and try to write down what its principles might be, the resemblance between shadows and perspective images becomes even more evident. Let's imagine that there's a projection from a pinpoint light source, sitting a finite distance from the screen and from the object that casts a shadow. Shadow theory says that the shadow of a dot is a dot, that the shadow of a straight line contains the shadow of all the points on that line, and that the shadow of a straight line is a straight line (with one exception: if the line passes through the light source, then its shadow will be a dot). To build this shadow theory one must simply perform some rewriting, starting with the principles of the classic theory of perspective: substituting "shadow" for "perspective" and "light source" for "eye." In fact, the perspective image of a dot is a dot, and the perspective of a straight line contains the perspective of all points on that line, and the perspective of a straight line is a straight line (except for the straight line that passes through the eye), and so on.

This brings us back to painting and its history. Shadows are the cheapest example of a projection. They can be found just about anywhere, and their construction is guaranteed by the way our world works. We can't go wrong in projecting a shadow. Perspective images follow the same laws of projection, but they need to be built. It's very easy to make a mistake with an image. (The camera obscura partly solves this problem.) For this reason—more than for any implicit understanding of the laws of perspective—it's plausible to think that shadows served as a model for the discovery of perspective.

Possible and Impossible Shadows

So in the end, shadows and perspective images go back to the same mathematical structure. Which means that we can examine the math of shadow as a particular example of a more general phenomenon. Let's go back to perception to see what this is about.

Shadows *look like* the objects that cast them. There are some moments or situations where a shadow is extremely similar to the thing casting it. It's fun to play with shadows in the late afternoon or early morning, when the sun is low in the sky and the shadow isn't squashed—isn't a puddle of ink, doesn't cancel out its owner's features. But even when a shadow undergoes remarkable distortion, we may still recognize the shape of the thing casting the shadow. First of all, it's hard to mistake the attribution of a shadow to its owner just by looking at its shape. A mere glance at two different shadows is enough to decide which one comes from a rabbit and which one from a hand—unless the hand is playing some shadow game and imitating a rabbit.

This ability to recognize things from their shadows is based on the ability to recognize similar shapes. We have no problem associating figures and shadows.

A rabbit trying to make things even more complicated.

Which of the two figures on the left could be a shadow of each figure on the right? Do you even have to stop and think?

Interestingly, however, we don't hesitate even when the figures in question are very distorted. In short, we recognize that the circle looks more like the ellipse than the rectangle looks like the square. But in what sense does an ellipse look like a circle, or a rectangle like a square? In what sense does an ellipse look *more* like a circle than like a triangle? (A mathematician would actually say that a circle *is* an ellipse.)

To answer this question we have to consider possible and impossible shadows again, this time looking at them even more generally (looking, that is, at shadows and objects outside the spatial context that they're projected in, considering each of them as simple geometric forms). Our vision is sensitive to distortions, but it tries to trace which elements a figure has in common with its shadow. These common elements permit us to perceive the parallelogram of the figure above as the shadow of a square. Our vision interprets the shadow less as a parallelogram than as a *distorted square.*

If you think about it, we know a lot of other things about the abstract geometry of shadows. We know that the shadow of a square cannot be a triangle or a pentagon; it must be a shape with four sides. We know that a circle's shadow can be an ellipse, but it cannot be a triangle or a square. More simply, we know that the shadow of a square cannot be made of *two* separate squares. There's more to it than the number of sides. The shadow of a shape without any hollows cannot itself have hollows (the shadow of a decagon cannot be star-shaped).

So we have to ask: how do we know these things?

Knowing this means that we can "see" the most abstract properties of geometric figures and, furthermore, that we can see how these do (or don't) remain the same when they're projected. This is a very complex skill that shows how our cognitive systems have shrewdly assimilated the math of projections.

How to Straighten a Slanted Shadow

Take a business card and hold it under a lighted lamp above a flat table. Its shadow will almost always be a bit slanted (at least two of the shadow's opposite sides will converge). Here's a puzzle: *how can you straighten a slanted shadow*—how can you make a shadow with parallel sides? There are two simple solutions. One consists of holding the card horizontally. The other consists of bringing the card and the table out to the sunshine (without worrying about their alignment). Try it and you'll see. In the first case we adjusted the tilt of the card in relation to the table. In the second case we adjusted the projection rays (assuming that sun rays are practically parallel, because they pour down from such a distant source). In both cases we *modified the method of projection.* One last solution combines the first two: we stand in the sun and we keep the tabletop and the card parallel. Now the shadow is a perfect copy of the original card.

The vicissitudes of slanted shadows allow us to classify projection methods quite simply. We have to keep two factors in mind. First of all, we must see whether the plane of the object is parallel to the plane where the shadow is projected (for example, a horizontal card above a horizontal table). Second, we have to ascertain whether the projection rays are parallel or convergent (that is, whether the light source is like a lamp or like the sun—practically an infinite distance away).

Moving from one projection method to another has an effect on the kinds of properties that the card shares with its shadow. This becomes evident when we review the card-and-table sequence backward. If the rays are parallel and the planes are parallel, then *all* the properties of the card's shape are passed on to the shadow: the shadow *copies* the length of the sides, the fact that they meet at right angles, and so on. This is called a *metric projection* because it maintains all the distances between any two points in the original. There is no distortion. Staying in the sun, but tilting the card in relation to the table, we get a shadow that copies some

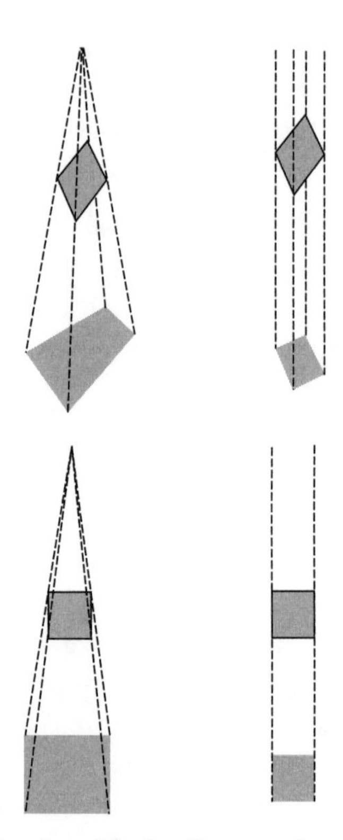

At top left is a slanted shadow. Two ways of straightening it: take the card out to the sunshine (top right), or hold it so that at least one of its sides is parallel to the table (bottom left). At bottom right is a way to make the shadow look like the card: take it out to the sunshine and also hold it parallel to the table.

but not all of the properties. The rectangle is copied as a parallelogram. The fact remains that the sides are parallel, but their lengths and the angles at which they meet cannot copy the length of the card's sides and the angles of the card's corners. This is an *affine projection.*

When we go back indoors and switch on the lamp, we return to the kingdom of central projections (including perspective projections). If we hold the card horizontally like the table, we create a shadow that copies the angles of the card and keeps the sides parallel, but lengthens them (the closer the card is to the light source, the longer they are). This projection is a *similitude.* The shadow is exactly like the card, but bigger.

The least faithful projection method is the one we started with, where the card is tilted beneath a lamp. Here we get the most distortion. But this is also the most general, so this kind of projection is pleonastically called a *projective projection.* This is the mother of all projections; all the others can be considered varieties of this one.

Our vision is sensitive to these mathematical properties; we are able to recognize what the shadow copies from the object that casts the shadow. Naturally, it would be implausible to think that we developed such a specialized cognitive skill only in order to recognize the geometric properties of shadows. One hypothesis that I find reasonable is that our recognition and understanding of shadows is like a parasite on our other perception skills. For example, the brain must have long ago solved the problem of treating certain two-dimensional configurations (the distribution of stimuli on the retina, which is practically flat) as *aspects* of structures that are really three-dimensional. Take a snapshot of your house and clip out the area that corresponds to the rug. This will almost always be a slanted trapezoid. But when you look at the carpet, or the photograph, it feels like you're seeing a rectangle. Your brain straightens out slanted shapes. By contacting the Straightener-of-Slanted-Shapes Department, the brain understands the structure of shadows—even though it cannot literally straighten them out for your eyes.

Your brain's geometric skills are used for understanding shadows. And from a certain point in the history of Western painting, drawing shadows became just a simple mathematical routine.

If It's Outside the Frame, It Doesn't Exist

The history of shadows in painting moved rapidly; shadows were eventually standardized in a set of fixed rules that became the province of fine-arts courses. But even so, shadows went on being unwelcome guests in paintings. Even when painters learned how to construct them and started loving to draw them—even when artists had access to tools like the camera obscura—sometimes shadows fell by the wayside. Maybe that's what happened with Bernardo Bellotto (1721–1780), known as Canaletto (he was the nephew and student of Antonio Canal, who had the same nickname). His great genre painting, *The Imperial Summer Residence of Schönbrunn Seen from the Courtyard,* documents an actual historical event. On August 16, 1759, Maria Theresa of Austria was awaiting Count Kinsky, who brought word of the defeat of Prussia's Frederick the Great. This was a big event, but we care only about the small details here. On the ground at the left, the shadow of a decorative vase seems to be missing. It's true that the vase itself doesn't appear in the painting, but the symmetrical architecture calls for one to exist just outside the painting's border. This leads one to think that the shadows were drawn separately and that Bellotto, or an assistant, found himself with the embarrassing problem of whether to add an extra shadow (if the vase didn't really exist) or to omit a shadow that should be there (if the vase was part of the architecture). He made the conservative choice and omitted the shadow. But in doing so he also eliminated the witness to the object that produced it. The painting is memorable less for the characters it shows than for its unusual content: it declares that a certain object located outside the painted scene does not exist.

In Bernardo Bellotto's Imperial Summer Residence of Schönbrunn Seen from the Courtyard, *the foreground shadow is incomplete. Shouldn't there be a shadow of a decorative vase located outside the picture frame?*

XVIII Memories from the Hereafter

Out, out brief candle!
Life's but a walking shadow.
—Shakespeare, *Macbeth*

One Last Story

"I could have called this purely mathematical work of mine *Memoirs from the Hereafter*, because it is the fruit of the reflections of a young lieutenant of engineers who was given up for dead and left on the battlefield of Krasnoe, near Smolensk, and long erased from the lists of the French army. During the horrible retreat from Moscow, 7,000 Frenchmen, exhausted by hunger, cold, and fatigue, found themselves launching an offensive there, without any artillery, on the orders of Marshal Ney, on November 18, 1812—St. Michael's Day in Russia; this final raging assault went against 25,000 fresh, well-equipped soldiers armed with forty guns led by Marshal Prince Miloradovich (who was soon to succumb to a military plot). . . ."

Few math books can boast an opening like this. And that's only the beginning. The narrator is an old professor, Jean-Victor Poncelet, embittered by a lifetime of discoveries for which he got no credit; by jealous colleagues; by the annoying administrative tasks that took time away from his research. His memoir reaches fifty years back to the events of the retreat from Russia. Poncelet studied at the Ecole Polytechnique and became an army officer of engineers. Called up during the Russian campaign, he was stationed in Smolensk while Napoleon's army marched toward Moscow. On November 9, 1812, Napoleon was retreating and he stopped in Smolensk long enough to grasp that he couldn't make winter camp there. Marshal Ney gathered a rearguard that left the city on the seventeenth; he hoped to meet up with the emperor on the eighteenth at Krasnoe; he was misinformed and he ran into Miloradovich instead. The surprise was mutual, the battle brief and violent. Ney was forced to fall back toward Smolensk; only 3,000 men from his battalion managed to follow him; Lieutenant Poncelet was among the prisoners taken on December 18, one of the darkest days of the retreat. This was the start of his odyssey. From Krasnoe the twenty-four-year-old lieutenant was sent to Saratov on the banks of the Volga. As the tatters of La Grande Armée crossed into France, the prisoner headed east, covering more than 600 miles in three months of forced marching through the Russian winter.

For the Fatherland, for the Sciences, and for Glory is the Napoleonic motto that can still be read on the façade of the old building of the Ecole Polytechnique in the center of Paris. When Jean-Victor Poncelet stepped into the fortress of Saratov, his dreams of glory were fading and the fatherland could not have seemed farther away. The exhaustion of the march suddenly caught up with him and he fell ill. Recovery was slow. Because his prison term promised to be long, Poncelet decided to dedicate his time to science. This was a tough task in primitive and isolated Saratov. Furthermore, his school memories had faded during his two years of fighting. But Poncelet patiently tried to review all the math he had learned, starting from the beginning. What happens in a situation like this is that you rediscover the basic mathematical propositions, and if you're lucky, you come across new ones. Something even better happened to Poncelet in the strange mental liberty of prison: he rewrote geometry, taking it to a new level of generality that had been only vaguely anticipated by his predecessors (great though they were), and he laid the foundation for modern geometry. For example, he discovered the principle of duality. (Consider this sentence: *Two lines meet in a point.* Now switch the terms "point" and "line" and change the verb to "determine"; you get *Two points determine a line.* Points and lines, at a deeper level, are examples of one and the same reality.) This is a fundamental principle that, with a simple rewrite (like the swoop of a magic wand), immediately multiplies the number of known theorems.

One biography claims that, in the absence of reference books at Saratov prison, Poncelet's key idea was to study the most general properties that shapes share with their shadows. I haven't found any confirmation of this story. Although I have tried to debunk numerous myths in this book, I like this story because it elegantly sums up and explains Poncelet's great scientific contribution.

Poncelet's innovation was to construct a geometric system in which shapes are classified on the basis of their projective properties. In Poncelet's geometry a shape's projective property is a property that is *invariant* in relation to a central projection. We saw that the shadow of a square can be a parallelogram; the distance between the corners and the angle of the corners are not, therefore, projective properties of a square. But the number of corners is: a square has four corners and so does its shadow. From the standpoint of projective geometry, a square and a parallelogram are the same thing. If you put yourself where the lamp is, you cannot tell the difference between a square and its shadow, whatever shape the shadow takes when it hits surfaces with different slopes.

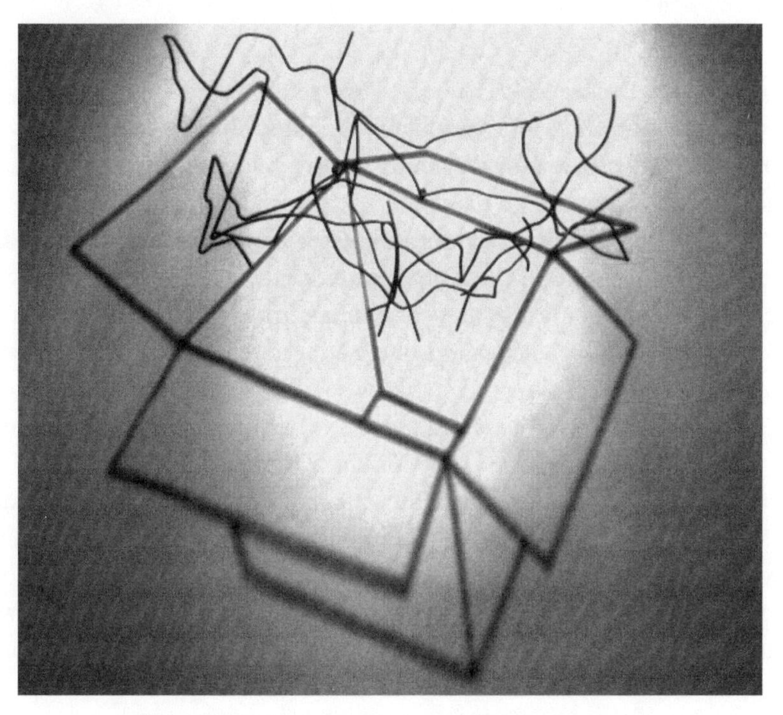

Projection unveils the hidden properties of shapes. Because the wire seems so messy, its tidy shadow (the box) comes as a surprise. Implausible Container, *a sculpture by Larry Kagan.*

The shadows bring us back to Desargues, whose math is presented by Poncelet as being similar to his own. There is one case where Desargues's Theorem seems not to work: when a triangle and its shadow have parallel sides, one cannot find the meeting points of the extensions of their sides on the plane.

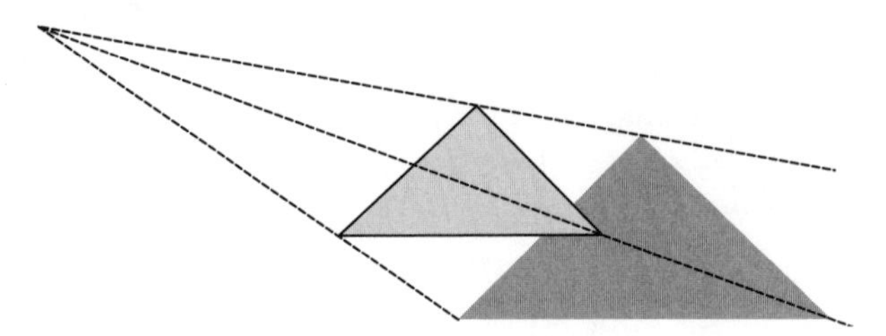

A special circumstance for Desargues's Theorem: if the sides of the two triangles are parallel, then one triangle can be the shadow of the other. But the extensions of their sides meet in infinity.

To salvage Desargues's Theorem, Poncelet postulated the existence of *infinite points,* which are completely counterintuitive but are now a common part of geometry teaching (many readers will recall learning the strange sentence "Two parallel lines meet at an infinite point").

Poncelet's biography seems to contain all the clichés of scientific mythology. There are the usual premonitions. As a boy, he supposedly oriented himself in the darkness of an abandoned mine thanks to a map that he drew as he walked, and so he managed to save a group of fellow explorers. He had the predictable weak health, the delayed schooling, the immersion in the classic books (at an age when he had barely learned to read), the long nights of study, the enrollment in a distinguished high school, and finally the entry into the Ecole Polytechnique. Add to this the story of his imprisonment, the scanty material he had to work with when captive, and even his mental slate wiped clean of his university learning (all these were weapons that Poncelet used in battling with his colleagues to claim primacy in certain discoveries, such as the principle of duality). Anecdotes too good to be true, tailored to burnish the aura around the invention of the new geometry.

But maybe this is all right, at least as far as our story is concerned. In the Saratov Prison, Poncelet nailed down the mathematical structure of shadows, guaranteeing them the legitimacy denied them for so many centuries. The vicissitudes of the European wars freed Poncelet in 1814.

Returning to France, he brought with him the notebooks that were the basis for the *Treatise on the Projective Properties of Figures.* With its publication, shadows finally came out of hiding with their heads held high. But there was a price to pay for coming out of Plato's prison: their definitive reduction to mere abstract, geometrical bodies.

FINALE

(Curtain rises)

Plato and His Shadow

At the edge of the sea. Plato looks at Skia, his shadow, as she stretches up the walls of the houses. The light is softening as the sun sinks into the sea.

Skia: So let's wrap this up before night comes.
Plato: What is to become of you?
Skia: You can finally get rid of me. In a few minutes I'll disappear into the darkness of the earth. However, I don't really care at all.
Plato: You're so fleeting that I couldn't even hurt you if I tried. But don't you really have any feelings? I'm sorry that we have to part like this.
Skia: Dear Plato, I don't know sorrow because I have no memory. That's just the way I am. My name, Skia, also means *trace,* but you won't find any traces on me—nothing like the scar on your arm that reminds everyone that you hurt yourself when you were a kid. I have no stories to tell because I don't carry any marks of the past. But, see, this is also a benefit for me. I'm always different, but at *each* moment I'm obliged to be the exact image of whatever I'm the shadow of. This is why geometers, astronomers, and painters have always trusted me. Since I have no memory, I can't play tricks when I deliver any message entrusted to me. What I say is above all suspicion.
Plato: Now I think I understand why shadows are so important. They're really extraordinary things.
Skia: They're extraordinary because they are halfway between thought and perception.
Plato: What do you mean?

SKIA: Every shadow carries a message hidden within its dark folds. Shadows are full of ideas. But they're ideas that everyone can see.

PLATO: Sort of like a word, if the word is written in a language that you know. This is why scientists have been able to talk with shadows.

SKIA: And I'll tell you something else. Scientists are able to talk with shadows because every shadow *is itself* a scientist. Because, for example, a shadow can build a two-dimensional model of a real physical thing. And it does it, tirelessly, all the time.

PLATO: But what's the use of that?

SKIA: Have you forgotten about Eratosthenes? He examined a little tiny shadow in the bottom of a bowl and he figured out how big the earth was!

PLATO: I've learned my lesson. If I were to write another book about knowledge, I would treat you with more respect.

SKIA: The sun is about to go down. We can't undo what's been done, but let's enjoy this beautiful sunset before I slip into the great shadow of the night.

The sun sinks into the sea. Skia breaks away from Plato's body and flies off silently. She hurtles over the houses and stretches across the mountains behind Athens.

The Discovery of Shadow

O empty shadows, save in aspect only!
—Dante, *Purgatorio,* II, 79

Who gets the last word? Plato or his shadow?

The mind has an uneasy relationship with shadows. The brain always uses shadows very shrewdly in order to learn about objects and where they are in the surrounding world. And yet the brain cannot quite grasp shadows. They're strange and confusing. Why is there so much ambiguity about the knowledge of shadows?

We must draw a distinction between the automatic, unconscious use of shadows, and the conscious use that requires the user to have some notion about them. This analogy comes to mind: I know how to use a tennis racket, but if you asked me to describe the racket's positions during a serve, I wouldn't know where to start. And when I'm actually serving, it's better if I don't think about the racket at all. When I'm playing, I hardly even notice the racket: I'm thinking about the ball and about the spots on the court where I want to send the ball.

Shadows are a bit like that racket. *We don't usually notice shadows.* When our visual system is turned on and working normally, we see trees, and chairs, and animals running: we ignore the shadows. We know that information about shadows is registered somewhere in the brain because, if it weren't for shadows, objects would seem to hover in midair and lose their solidity. But we must explicitly pay attention to shadows in order to register them. This is interesting because shadows in themselves are quite visible: there's a sharp change in brightness wherever there's a shadow in your field of vision. Shadows do all they can to grab your attention. And yet in the end they are only walk-on players; they're just extras in the movie of your perception. Attention must be explicitly paid to shadows.

This observation goes together with what we know about the minds of small children. We don't come equipped at birth with an idea of shadows. Shadows are not part of that small trove of basic concepts that describe the world, that valuable little biological inheritance handed down by the animals we evolved from. One of these concepts is "material object" (which describes rocks, obstacles, things that can't move by themselves); another is "living object" (things capable of autonomous movement); these apparently belong to that biological inheritance. From an evolutionary point of view, it's fairly reasonable. Your chances of survival improve if you don't need to *learn* (because you already know) that you have to avoid things that are obstacles, and that there are other things that might hunt you down. But shadows don't seem important enough to deserve description by an innate theory or concept.

One simple hypothesis is that we *discover* shadows as children; we find them *surprising;* and we cobble together a somewhat gerrymandered theory (which, as we've seen, causes problems when we think about shadows even as adults).

Children's metaphysics—and the metaphysics that underlies many of our adult intuitions—is based on three major principles: an object cannot be in two places at one time; two objects cannot be in the same place at the same time; and if an object moves, it follows a path without any gaps—it cannot disappear and then reappear. Cartoonists know this and they entertain us (children and adults alike) with characters who split into two, who overlap, disappear, and reappear. In everyday life there are some phenomena that violate these principles. An object reflected in the mirror looks like a duplicate object. The beam of a flashlight can be broken for a moment by some object and then continue farther on. Shadows intersect and overlap in space. All these are magical objects.

Van de Walle, Rubenstein, and Spelke's experiment showed that little children get confused by shadows, and that they try to apply to shadows the principles that suit only material objects. For example, they're not surprised (as they should be) when a shadow moves along with the box it falls on—perhaps because the shadow is touching the box, and two things that touch ought to move along together. Newborns don't have a prefabricated idea of shadows. Maybe older children are attracted to shadows precisely because they don't know how to classify them; they see them as metaphysical pirates who violate all the known laws of the physical world. The strange but not completely irrational architecture of the concept of shadows is probably born of the surprising nature of shadows, and of the attempt to understand them.

Examining the adult concept of shadow, we find certain characteristics cropping up like cognitive fossils. We can dig these up to understand how certain intuitions came together into a small theory.

The concept of shadow is a *spatial* concept; shadows are (dark) areas. But space comes into it subtly. It has a peculiar role because the concept of shadow is partly *figural* and partly *causal.* We often try to define shadow as the area where light has been blocked off; but as we've seen, night fits this definition and yet we don't think of night as shadow. We have also seen that an object cannot cast a shadow through another object. This is because the first object's shadow gets lost in the second object's shadow. We tend to use the term "shadow" only where we see a border between light and shadow; that's why the concept is *figural.* Shadow needs to have a figure that we can see, or at least imagine. The shadow figure is unusual because things that have a figure are generally material. A hole is one exception to this rule; but holes are also tactile objects, while shadows are only visual. Shadows are pure figures.

Furthermore, we usually think of shadows as things that are *projected.* This is why we don't accept that shadows can pass through other objects. If it meets a screen, the shadow stops. The causal component of the shadow concept—which allows us to see a shadow as something that can be thrown against an obstacle—also means that we grant shadows the typical impenetrability of things that can be cast. Shadows may be able to interpenetrate, but we don't know what happens when they interact with material objects.

We discover that shadows are *products* partly because we learn to produce them. Causality in the concept of shadow is linked to an image of ourselves as agents capable of changing the world around us. We can control shadows with our movements, just as we control the tennis racket. Yet—and this is an interesting revelation for children—we control them only partway: for example, we cannot make them spin around us. This might explain why shadow easily becomes a vehicle for magnificent psychological images. Shadows are our slaves, but they don't obey us completely; they can revolt, displaying a will of their own, revealing a soul.

Using shadows in perception doesn't require that one has a concept of shadows, but such a concept is needed for *thinking* about shadows. This concept is developed—and it functions—by assembling resources from various parts of the brain: it's a real mental parasite. But there is some justice in this, because shadows are metaphysical pirates. Part of the appeal of science—and the normative spirits subjugated by Plato failed to notice just *how* true this is—is that even pirates and parasites are welcomed to the field as long as they do their job well.

APPENDIX: SHADOW NEWS, SHADOW FACTS

The Names for Shadow

When we think about shadow, we generally think about the silhouette that is cast on a surface, the dark shape that seems to mimic the object of which it's the shadow. In this book I often use the word "shadow" to discuss this kind of shadow, but also as a name for other shadow phenomena. Whenever the context is unclear, it's useful to employ slightly more technical terminology derived partly from painting and partly from architectonic drafting. The narrowest sense of shadow—the figure—is called the *cast* shadow. The darkness on the dim side of an object is called *self-shadow* (painters know it as *shading*). The lines separating shadow from light are called *shadow lines* or *terminators.* The *shadow zone* is all the space that gets no light because the object is hiding the source of light. A *silhouette* is what we can see of a body that comes between us and the light. Here is a sketch using these concepts. The light source (ideally, a single point located at *a*) illuminates a sphere that casts a shadow onto a screen.

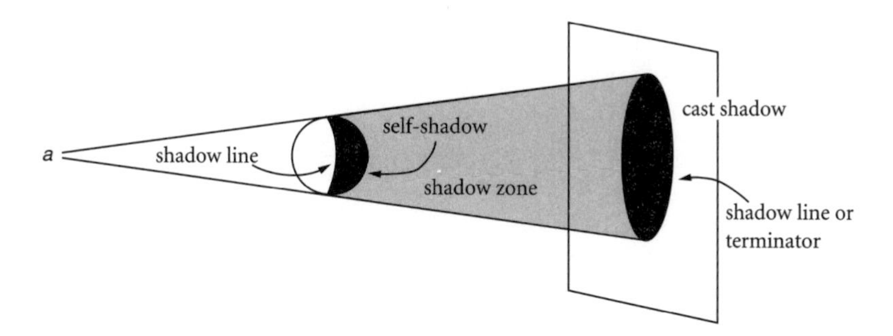

If an observer was in the shadow zone, he would see the sphere as a black circle—the silhouette of the sphere. (X-rays show the silhouette of the bones and the internal organs.)

This diagram shows an ideal light source: it's just a single point of light. Reality is more complicated, because most light sources aren't just one dot. But a source that's not a single point—such as the sun, a candle flame, or the filament of a bulb—can be pictured as the sum of several single-point lights, each of them casting a shadow that's slightly offset from the others. The *penumbra* is the region reached by light from only some of these light sources; in this case the word "shadow" acquires a new meaning as the region untouched by light from any of these sources.

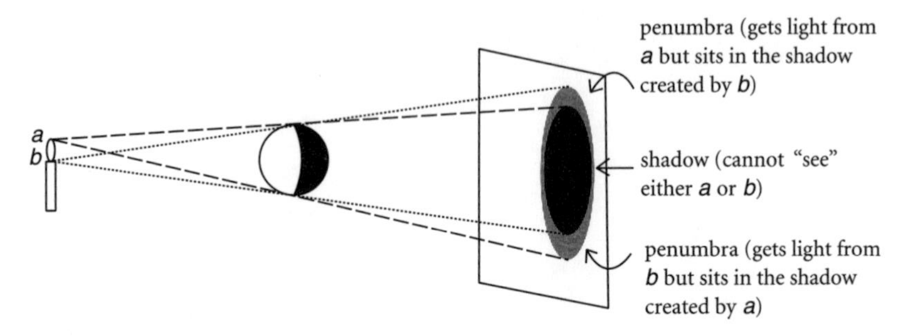

The shading-off that you see around a shadow's edges is actually the penumbra effect.

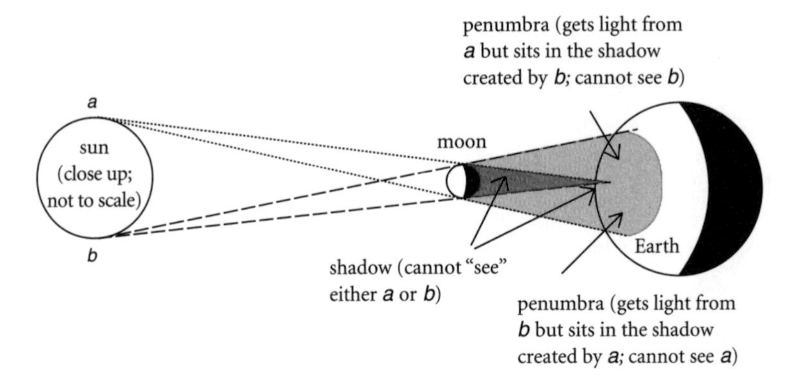

Above are the names for the parts of a solar eclipse; the moon hides the sun and casts a shadow on the earth. (The sun and moon are shown here as being closer than they are; they're not drawn to scale.)

Chief Discoveries and Scientific Measurements Regarding Shadows, or in Which Shadows Were Instrumental

> All celestial observations are made by means of light and shadow.
>
> —Kepler, *Ad Vitellionem Paralipomena*

- date unknown: Night is the shadow of the earth.
- date unknown: The moon has a cycle of phases.
- date unknown: The moon is a sphere, as seen from the geometric evolution of its phases.
- date unknown: Lunar eclipses are caused by the earth's shadow; solar eclipses are caused by the moon coming between the earth and the sun. The solar corona can be seen during a solar eclipse.
- date unknown: The earth is not flat, as can be seen from the differing lengths of shadow at different latitudes at noon on the same day.
- date unknown: The shortening and lengthening of shadows from one solstice to another shows that the sun's trail through the sky is inclined in relation to that of the stars; this accounts for the seasons.
- date unknown; traditionally attributed first to Parmenides, who was active around 500 B.C.: The moon doesn't really wax and wane, but rather is lighted by the sun. The moon seems to "look at" the sun.
- date unknown; traditionally attributed first to Anaxagoras, born c. 500 B.C.: The moon is lighted by the sun.
- Aristotle (384–322 B.C.): The earth's shadow on the moon during an eclipse proves that the earth is spherical and larger than the moon.
- Aristarchus of Samos (third century B.C.): The distance between the earth and the sun is much greater than the distance between the earth and the moon. Two shadow-based methods (the method of dycothomy and the method of the eclipse diagram) are found for measuring the dimensions of the bodies and the Earth-to-moon and Earth-to-sun distances. These methods were geometrically rigorous but depended on readings that were too approximate; however, they were to be used more or less successfully for two thousand years.
- Aristarchus: The shortening and lengthening of shadows can also be explained by the hypothesis that the axis of rotation of the earth is inclined in relation to the axis of its revolution around the sun. Seasons are a shadow phenomenon linked to the spherical nature of the earth and the inclination of its axis.
- Eratosthenes (c. 273–192 B.C.): Comparison of shadows at two points on the same meridian allow one to determine that the circumference of the earth is 250,000 stadia.
- Hipparchus (second half of the second century B.C.): The moon lies an average of 67⅓ terrestrial radii away from the earth.
- Hipparchus: The sun lies an average of at least 490 terrestrial radii away from the earth.

- Theon of Alexandria (end of the fourth century A.D.): Shadow geometry should demonstrate that light travels in a straight line.
- Galileo, 1610: The shadows cast by the rising sun on the moon reveal a rough surface. Therefore, the moon is not a (very) different type of object from the earth.
- Galileo, 1610: The mountains of the moon are almost 5 miles high.
- Galileo, 1610: Venus does not give off light of its own.
- Galileo, 1610: The cycle of Venus's phases is incompatible with Ptolemaic theory. Venus must revolve around the sun, so the sun is just one of the universe's three centers of rotation (the others being Earth and Jupiter, whose satellites were discovered by Galileo).
- Galileo, 1612: The eclipses of Jupiter's satellites constitute a great cosmic clock useful in determining longitudes on Earth.
- Gassendi, 1631: The silhouette of Mercury, observed during that planet's passage across the sun, demonstrates that Mercury is only one-sixth the size it was formerly thought to be.
- Horrocks, 1629: Observation of the transit of Venus, with results similar to those found by Gassendi for Mercury.
- Campani, 1664 (and other astronomers in the same decade): Shadows show that Saturn's bizarre shape is due to the rings circling it, as hypothesized by Huygens in 1656.
- Grimaldi, 1665: Shadows show that light spreads not only in a straight line—by reflection or by refraction—but also by diffraction.
- Rømer, 1676: Delays in the eclipses of Jupiter's satellites, observed at different points in the year, demonstrate that light travels at a finite speed and make it possible to calculate that speed.
- Halley, 1715: First precise observation of the path of an eclipse, made possible by the publication of a prediction made by Halley himself.
- 1761: The transit of Venus (the first one since 1639) allows the first accurate measurement of the solar parallax.
- Eddington, 1919: The solar eclipse of May 29 permits the observation of a star that appears slightly out of place in the constellation of Taurus, where the sun is passing through. The deviation of the light by the solar mass accords with the predictions of Einstein's theory of relativity. Various other observations and discoveries were made possible by solar eclipses in previous years: for example, spectrographic analysis on August 18, 1868, allowed Joseph Norman Lockyer to discover helium, which was isolated on Earth only in 1895.
- Dunham, 1980: A discrepancy between the actual and the theoretical path of the eclipse of April 22, 1715 (registered by Halley) leads to the hypothesis that the radius of the sun has shrunk by ⅓ of 1" of an arc in the past three centuries.
- Stephenson and Morrison, 1984: The discrepancy between the actual and the theoretical paths of about seven hundred historic eclipses leads to the hypothesis that the earth's daily rotation has slowed by about 1/20 of a second in the past 2,500 years.

BIBLIOGRAPHIC NOTES

I *In the Beginning There Was Shadow*

The myth of the cave: Plato, *Republic*, bk. VII. A philosophical reevaluation of shadows, in S. Todes, "Shadows in Knowledge: Plato's Misunderstanding of Shadows and of Knowledge as Shadow-Free," in *Selected Studies in Phenomenology and Existential Philosophy*, ed. D. Ihde and R. M. Zaner (The Hague: Martinus Nijhoff, 1975), pp. 94–113.

The sun doesn't see shadows: Goethe, *Faust*, pt. II, act 3.

Galileo talks about the blackened statue in a letter to the painter Cigoli of June 26, 1612.

Shadows and makeup: B. Horn, "Obtaining Shape from Shading Information," in P. H. Winston, ed., *The Psychology of Computer Vision* (New York: McGraw-Hill, 1975), pp. 115–155; this is a classic study of the reconstruction of surface slope on the basis of shading. In the same volume, see D. Waltz, "Understanding Line Drawings of Scenes with Shadows," pp. 19–91, for an analysis of simplified scenes in which shadows help in understanding objects' shapes.

Experiments with shadows and wire objects: H. Wallach and D. N. O'Connell, "The Kinetic Depth Effect," in *Journal of Experimental Psychology*, 45, 1953, pp. 205–217.

For a history of light that ends with unexpected praise for shadow: David Park, *The Fire Within the Eye* (Princeton, N.J.: Princeton University Press, 1999).

II *Ancient and Modern Shadows*

Epigraph: From A. B. Yehoshua, *Mr. Mani* (New York: Doubleday, 1992).

Shadows over Tokyo: Lester Thurow, "Asia: The Collapse and the Cure," in *New York Review of Books*, February 5, 1998, p. 23.

Shadows in Japanese houses: Junichiro Tanizaki, *In Praise of Shadows* (Stony Creek, Conn.: Leete's Island Books, 1988).

I chose from a vast literature on shadow theater: P. Kahle, "The Arabic Shadow Play in Egypt," in *Journal of the Royal Asiatic Society*, 1940, pp. 21–34; L. K. Myrsiades, *The Karagiozis Heroic Performance in Greek Shadow Theater* (Hannover and London: University Press of England); J. Cuisinier, *Le Théatre d'Ombres à Kelantan* (Paris: Gallimard, 1957); G. Jacob, *Geschichte des Schattentheater im Morgen und*

Abendland (Hannover: Orient-Buchandlung Heinz Lafaire, 1925); *Das Schattentheater* (Berlin: Mayer & Müller, 1901).

The Western fascination with Middle Eastern shadow theater is documented by G. Flaubert, *Voyage en Orient* (1848) (Paris: Belles Lettres, 1948), and by Gérard de Nerval, *Voyage en Orient* (1850) (Paris: Le Divan, 1927), in the chapter "Theaters and Festivals."

III Shadow Merchants

A study of the use of shadow metaphors in language and literary creation: J. Novakova, *Umbra* (Berlin: Akademie Verlag, 1964).

J. Frazer, *The Golden Bough* (London: Macmillan, 1922), chap. 18. The text uses data from an article by J. von Negelein, "Bild, Spiegel, und Schatten im Volksglauben, in *Archiv für Religionswissenschaft,* 5 (1902), pp. 1–37, which mentions the inherent contradictions in the metaphysics of shadows. The idea of shadow as soul is also discussed by L. Lévy-Bruhl in *L'Ame Primitive* (Paris: Alcan, 1927), p. 161ff., which notes that "shadow" and "soul" are themselves ambiguous terms, "the source of many errors." In English: Lucien Lévy-Bruhl, *The "Soul" of the Primitive* (New York: Macmillan, 1928).

The reports on central African populations quoted in the text are found in Germaine Dieterlen, *La Notion de Personne en Afrique Noire* (Paris: C.N.R.S. and L'Harmattan, 1973). There doesn't seem to be any monographic publication on the ethnography of shadows.

For persuasive evidence that in classical times people used the same images of shadow that are found in central Africa today: P. W. van der Horst, "Der Schatten im Hellenistischen Volksglauben," in *Studies in Hellenistic Religion,* ed. M. J. Vermaseren (Leyden: E. J. Brill, 1979), pp. 27–36. For the Egyptian world, B. George, *Zu den Altägyptischen Vorstellung vom Schatten als Seele* (Bonn: Habelt, 1970); here, among other information, is the fact that the hieroglyph for shadow is an umbrella: ( )

Transylvanian shadow merchants: Frazer, op. cit., p. 233.

On counterintuitive beliefs: Dan Sperber, *La Contagion des Idées* (Paris: Odile Jacob, 1997). In English: *Explaining Culture* (Oxford: Blackwell, 1996).

IV Shadows on the Mind

A very detailed work on children's concepts: S. Carey, *Conceptual Change in Childhood* (Cambridge: MIT Press, 1986).

Piaget often wrote about shadows; for example, in *The Child's Conception of Physical Causality* (London: Routledge, 1951), and (with B. Inhelder) in *The Child's Conception of Space* (New York: Norton, 1967).

New experiments in the spirit of Piaget: R. DeVries, "Children's Conceptions of Shadow Phenomena," *Genetic Psychology Monographs* 112 (1986), pp. 479–530.

On the distinction between material and immaterial: S. Carey, "Knowledge Acquisition: Enrichment or Conceptual Change?" in *The Epigenesis of Mind,* ed. S. Carey and R. Gelman (Hillsdale, N.J.: Erlbaum, 1991), pp. 257–291, on p. 279.

Habituation method: R. L. Fantz, "The Origins of Form Perception," *Scientific American* 204 (1961), pp. 66–72. Arithmetic in newborns: K. Wynn, "Addition and Subtraction in Infants," *Nature* 358, pp. 749–750; S. Dehaene, *La Bosse des Maths* (Paris: Odile Jacob, 1997). In English: *The Number Sense* (New York and Oxford: Oxford University Press, 1997).

Newborns in a world of objects: E. S. Spelke and G. A. Van de Walle, "Perceiving and Reasoning About Objects: Insights from Infants," in *Spatial Representation,* ed. N. Eilan, R. McCarthy, and B. Brewer (Oxford: Blackwell, 1993), pp. 132–161; E. S. Spelke, "Principles of Object Perception," *Cognitive Science* 14 (1990), pp. 29–56.

Newborns and shadows: G. A. Van de Walle, J. Rubenstein, and E. S. Spelke, "Infant Sensitivity to Shadow Motions," *Cognitive Development* 13 (1998), pp. 387–419.

V Shadow of a Doubt

Shadows are the most perfect two-dimensional items: E. A. Abbott, *Flatland* (1882).

Shadows and quantum physics: David Deutsch, in a chapter ("Shadows") from *The Fabric of Reality* (New York: Penguin, 1997), explains how the strange behavior of some shadows lets us deduce even the existence of parallel universes.

On the Great Shadow Brain-Teaser: S. Todes and C. Daniels, "Beyond the Doubt of a Shadow: A Phenomenological and Linguistic Analysis of Shadows," in *Selected Studies in Phenomenology and Existential Philosophy,* ed. D. Ihde and R. M. Zaner (The Hague: Martinus Nijhoff, 1975), pp. 203–216. The logician Bas van Fraassen took up the puzzle again in the context of an examination of the structure of scientific theories: *Laws and Symmetry* (Oxford: Clarendon Press, 1989), chap. 9.

R. Sorensen, "Seeing Intersecting Eclipses," *Journal of Philosophy,* 1999, pp. 25–49; he transfers the problem of double shadows to astronomy and to the philosophy of perception. Imagine two planets that move in between the earth and the sun and seem perfectly superimposed in the sky. Would we see the closer one or the farther one?

Things speedier than light: M. A. Rothman, "Things That Go Faster than Light," *Scientific American* 203 (1960), pp. 142–152; E. Sober, "A Plea for Pseudo-Processes," in *Pacific Philosophical Quarterly* 66 (1985), pp. 303–309.

On holes: R. Casati and A. Varzi, *Holes and Other Superficialities* (Cambridge: MIT Press, 1994).

On intuitive physics: P. Bozzi, *Fisica Ingenua* (Milan: Garzanti, 1991).

VI Special Effects

A collection of texts and fragments that refer to the astronomy of ancient Greece: T. Heath, *Greek Astronomy* (New York: AMS Press, 1969). Two fundamental readings on ancient astronomy are O. Neugebauer, *The Exact Sciences in Antiquity* (New York: Dover, 1969), and C. Walker, *Astronomy Before the Telescope* (London: British Museum Press, 1996).

Popper on Parmenides and the moon: *The World of Parmenides* (London: Routledge, 1998).

Aristotle thought that the moon didn't spin on its own axis: *De Caelo,* ii.8, 290a 26 (I used the edition published in Bari by Laterza, 1995).

On archaeoastronomy: A. Aveni, *Stairways to the Stars: Skywatching in Three Great Ancient Cultures* (New York: John Wiley, 1999); E. C. Krupp, *Echoes of the Ancient Skies: The Astronomy of Lost Civilizations* (Oxford: Oxford University Press, 1994). Cognitive models of the cosmos: S. Vosniadou and W. F. Brewer, "Mental Models of the Earth: A Study of Conceptual Change in Childhood," *Cognitive Psychology* 24 (1992), pp. 535–585.

The calendar bone: A. Marshack, "The Taï Plaque and Calendrical Notation in the Upper Paleolithic," *Cambridge Archeological Journal* 1 (1991), pp. 25–61.

Concave/convex ambiguities (examined in the 1930s by the psychologist Kai von Fieandt) are among the many cases studied in the psychology of perception, which considers how our vision manages to reconstruct objects' shapes based on shading. K. von Fieandt, "Uber Sehen von Tiefengebilden bei Wechselnder Beleuchtungsrichtung" (Helsinki: University of Helsinki, 1938). Von Fieandt wrote an essay on the perception of shadow: "Das Phänomenologische Problem von Licht und Schatten," *Acta Psychologica* 6 (1949), pp. 337–357.

On Aristarchus: T. Heath, *Aristarchus of Samos: The Ancient Copernicus* (New York: Dover, 1981), mostly discusses Aristarchus's heliocentric theory. A passage in which the succession of the seasons is explained as a phenomenon of shadow can be found on p. 305.

On the dimensions of the cosmos: A. van Helden, *Measuring the Universe* (Chicago and London: University of Chicago Press, 1985).

VII Eclipses, Shadow Cones, and Pyramids

As in the preceding chapter, Heath and van Helden are the basic references about Greek astronomy and planetary measurements.

There is a great deal of easily accessible literature on the mechanisms of eclipses. A collection of references to eclipses of the past is in F. R. Stephenson, *Historical Eclipses and Earth's Rotation* (Cambridge: Cambridge University Press, 1996). The Old Testament on eclipses: Joel III: 30, 31. On the Takana: K. Hissin and A. Hahn, *Die Tacana* (Stuttgart: Kholhammer Verlag, 1961).

There is still only scanty material on the effect of eclipses on animals and people. For example, G. Francillon and P. Menget, eds., *Soleil Est Mort* (Nanterre: Laboratoire d'Ethnologie at de Sociologie Comparative, 1979), documents the results of an expedition to Africa to study the ecology of the eclipse of June 30, 1973. Apparently, many passenger pigeons in Great Britain got lost during the eclipse of August 11, 1999.

Aristotle and the use of the moon for determining the shape of the earth: *De Caelo*, ii.14, 297b.

On Thales: Herodotus, *History*, I, p. 74. A fragment published in 1987 reappraises Thales' position. The fragment cites an opinion from Aristarchus—that is, from an astronomer—that Thales was the first one to have understood that the sun can be screened by the moon only when the moon is new. Thales did not predict a solar eclipse, in this view, but did say when it was possible that one might occur. The new fragment regarding Thales is discussed in A. V. Lebedev, "Aristarchus of Samos on Thales' Theory of Eclipses," *Apeiron* 23 (1990), pp. 77–85.

The Crucifixion eclipse: Luke 23: 44–45.

Quintilian and eclipses: *Istituzione Oratoria*, bk. I, 10, pp. 46–49 (Milan: Mondadori, 1999). Eclipses and superstitions: I. Drealants, *Eclipses, Comètes, Autres Phénomènes Célestes et Tremblements de Terre au Moyen Age* (Louvain-la-Neuve: Presses Universitaires de Louvain, 1995); Plutarch, *Vita di Nicias.* Shadows were never kind to Nicias: in the battle lost to the Boeotians, his soldiers had the moon behind them, so they were shrouded in their own shadows and appeared to the enemy to be far fewer and less well armed than they actually were.

Diogenes Laertius on Thales: For example, in Colli, *La Sapienza Greca* (Milan: Adelphi, 1992), vol. II, p. 125, fragment 10 B 1 27. Thales sacrifices a cow: In Colli, op. cit., p. 123, fragment 10 B 1 24.

On Egyptian mathematics: Neugebauer, *The Exact Sciences in Antiquity*, which offers a critical reconsideration of Thales' supposed discoveries (chap. VI).

VIII The Stolen Sundial

Pliny and the theft of the sundial: *Natural History*, VII, pp. 214–215. Vitruvius and the catalogue of ancient solar clocks: *De Architectura*, bk. IX.

The calculation of the error using the solar clock from Catania: S. L. Gibbs, *Greek and Roman Sundials* (New Haven, Conn., and London: Yale University Press, 1976), p. 96; this reference work contains a complete list of known Greek and Roman solar clocks and a mathematical description of them.

The fragment of Eratosthenes' poem *Hermes* is reproduced in J. U. Powell, *Collectanea Alexandrina* (Chicago: Ares Publishers, 1981). Virgil copied this passage in his *Georgics*, I.

On the north: See Luigi de Anna, *Thule: Le Fonti e le Tradizioni* (Rimini: Il Cerchio, 1998). Peary's sad tale is described in the centennial issue (September 1988) of

National Geographic. Robert M. Bryce's book *Cook and Peary* (Mechanicsburg, Pa.: Stackpole Books, 1997) contains much documentary material on the "conquest" of the North Pole.

On solar clocks in general: René R. J. Rohr, *Cadrans Solaires* (Paris: Gauthier-Villars, 1965). In English: *Sundials: History, Theory, and Practice* (New York: Dover, 1996). The third volume of J. Needham, *Science and Civilization in China* (Cambridge: Cambridge University Press, 1959), describes the Chinese science of solar clocks.

IX In the Shadow of the Minaret

Epigraph: From the Koran, sura 6, verse 96.

There is only one copy of the manuscript of *The Exhaustive Treatise on Shadows,* in the library of Bankipore, India. The copyist notes that he finished his work in Dhu al-Hijja in the year 631—in other words, between August 28 and September 26, 1234. The work (composed two centuries earlier) is dedicated to Sheik Musaphir, an eminent citizen of Nishapur in Khurasan. Al-Biruni's text, *Kitab fi Ifrad al-Maqal fi amr al-Zilal,* was translated and commented upon by E. S. Kennedy, *The Exhaustive Treatise on Shadows* (Aleppo: Institute for the History of Arabic Science, 1976). Kennedy is also the author of the biography of al-Biruni in the *Dictionary of Scientific Biography.*

The *Treatise* is not the only book on shadows from the era; there was a real genre of this kind, probably aimed at an audience of cultivated muezzins. Al-Biruni cites the writings of Abu al-Hasan Thabit bin Qurra, who died in 901, a Syrian translator from Greek to Arabic, a bridge between late Greek culture and Islamic culture (the reference to *On the Determination of Lines Traced at the Borders of Shadow at the Horizon* appears on p. 14 of Kennedy's commentary). He also cites the *Book of Shadows* by Ibrahim bin Sinan (a relative of the former, who died in 946). But he doesn't mention the other book of shadows that has come down to us, *On the Formation of Shadows,* by his great contemporary al-Hazen (Ibn al-Haytham), translated into German by E. Wiedemann in "Über eine Schrift von Ibn al Haitam: Uber die Beschaffenheit der Schatten," in *Sitzungsberichte der Physikalisch-medizinisch Sozietät in Erlangen* 39 (1907), pp. 226–248, with an important discussion of penumbra and references to other Arabic works on shadow.

X Time Flies Out Through the Hole in a Shadow

On Danti's solar clock: F. Mancinelli and J. Casanovas, *La Torre dei Venti in Vaticano* (Rome: Libreria Editrice Vaticana, 1980); O. Gingerich, "The Tower of the Winds and the Gregorian Calendar," in *The Great Copernicus Chase* (Cambridge: Cambridge University Press, 1992), pp. 82–88. A good description is in G. Paltrinieri, *Meridiane e Orologi Solari d'Italia* (Bologna: L'Artiere, 1997).

On the role of Cassini's solar clock: A. van Helden, *Measuring the Universe* (Chicago and London: University of Chicago Press, 1985), p. 130ff.

XI Shadow Wars

The *Starry Messenger (Sidereus Nuncius)* is available in various editions. I quote from the paperback version published by Classici Ricciardi/Einaudi. Announcement of great things, p. 11; the moon seems to be two terrestrial radii away from the earth, p. 15; variation of light and dark of the moon, p. 39; analogy between the earth and the moon, and sunrise on the moon, p. 21; spots of light and shadow, p. 29; the Bohemian crater, p. 27.

When computer science began to pay attention to shadows, it used studies of the lunar surface: T. Rindfleisch, "Photometric Method for Lunar Topography," in *Photometric Engineering* 32 (1966), pp. 262–276.

The moon has no mountains: Ludovico delle Colombe, *Contro il Moto della Terra,* in the edition of Galileo's *Works* (*Opere,* edited by Favaro), III, pp. 250–290, particularly pp. 286–287.

Kepler on the moon's spots: *Dissertatio cum Nuncio Sidereo* (Conversation with the Sidereal Messenger); I read it as *Discussione col Nuncio Sidereo e Relazione sui Quattro Satelliti di Giove,* ed. E. Pasoli and G. Tabarroni (Turin: Bottega d'Erasmo, 1972).

Plutarch, *The Face of the Moon:* The pertinent passages are on pp. 89–91.

On Harriott and Galileo: T. F. Bloom, "Borrowed Perceptions: Harriot's Maps of the Moon," in *Journal for the History of Astronomy,* 1978, pp. 117–122. S. J. Edgerton, Jr., dedicated a chapter to Galileo's images in *The Heritage of Giotto's Geometry* (Ithaca, N.Y., and London: Cornell University Press, 1991), p. 223ff., following a thread originally teased out by E. Panofsky, who was the first to draw attention to Galileo's artistic development: Erwin Panofsky, "Galileo as a Critic of the Arts," in *Isis* 47 (1956), S. 3–15. The portrayal of the moon in the *Immacolata* in

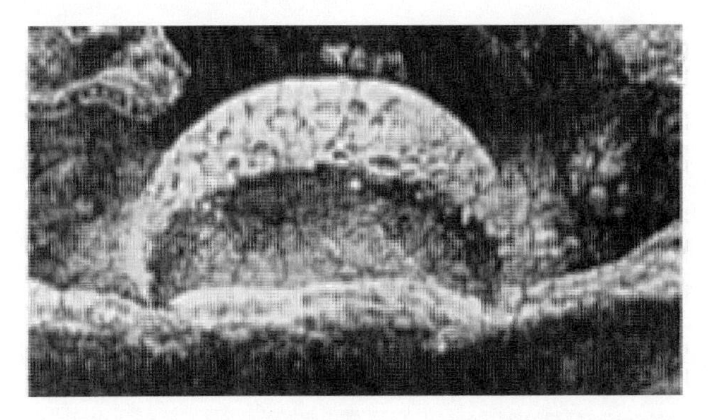

Ludovico Cardi, known as Il Cigoli, painted an Immaculate Virgin *in which the moon (detail above) represents purity, but after Galileo's astronomical discoveries, this simile turned out to be a double-edged sword.*

Santa Maria Maggiore in Rome deserves a chapter all to itself: it was executed by Il Cigoli after Galileo's discoveries (including, of course, the Bohemian crater). E. Reeves dedicated a book to this: *Painting the Heavens: Art and Science in the Age of Galileo* (Princeton, N.J.: Princeton University Press, 1997).

The identification of the Bohemian crater with the Albategnius crater: O. Gingerich, "Dissertatio cum Professore Righini et Sidereo Nuncio," in *Reason, Experiment, and Mysticism in the Scientific Revolution*, ed. M. L. Bonelli and W. R. Shea (New York: Science History Publications, 1975), pp. 59–76; E. A. Whitaker, "Galileo's Lunar Observations and the Dating of the Composition of 'Sidereus Nuncius,' " *Journal for the History of Astronomy* 9 (1978), pp. 155–169.

A powerful treatise on shadows came out in 1646: *Ars Magna Lucis et Umbrae* by the Jesuit Athanasius Kircher (1602–1680), a book about optics (light) and solar clocks (shadow). *Ars Magna* ranged from metaphysics (the universe is a mixture of light and shadow) to instructions for building toys (including a magic lantern).

XII Venus Imitates the Shadow of Diana

Kepler struggling with Galileo's anagrams: *Gesammelte Werke*, XVI (Munich: Beek, 1937), p. 357.

On the phases of Venus before Galileo's time: R. Ariew, "The Phases of Venus Before 1610," *Studies in the History and Philosophy of Science* 18 (1987), pp. 81–92.

Galileo on the book of the universe: L. Sosio, ed., *Il Saggiatore* (Milan: Feltrinelli, 1965), text 6, p. 38.

Gassendi and Mercury: *Mercurius in Sole Visus*, in *Opera Omnia* (Lyon, 1698).

On Horrocks: Entry in the *Dictionary of Scientific Biography*, edited by W. Applebaum.

On the whole period: See vol. 2A of the *General History of Astronomy*, ed. R. Taton and C. Wilson (Cambridge: Cambridge University Press, 1989), and van Helden, *Measuring the Universe*.

XIII Maybe Saturn Devoured His Own Children?

Galileo on Saturn devouring his children: Letter of December 1, 1612, to Marco Velseri, p. 174 in the Ricciardi edition of the *Sidereus Nuncius*.

The battle between Divini and Campani is recounted by A. van Helden and M. L. Righini Bonelli in *Divini and Campani: A Forgotten Chapter in the History of the Accademia del Cimento*, supplement to the *Annali dell'Istituto e Museo di Storia della Scienza* (Florence, 1981).

The *Cosmotheoros* speaks of Saturn on pp. 776–789 of the *Oeuvres Complètes de Christiaan Huygens*, Société Hollandaise des Sciences (The Hague: Martinus Nijhoff, 1888), vol. XXI. The incorrect drawing is on p. 125 of the English edition of 1698.

XIV The Speed of Shadow

An introduction to the problem of longitude: D. Sobel, *Longitude* (New York: Walker Publishing Co., 1995).

Al-Biruni on eclipses is quoted in F. R. Stephenson and S. S. Said, "Precision of Medieval Islamic Eclipse Measurements," *Journal for the History of Astronomy* 22 (1991), pp. 195–207, on p. 196.

Using eclipses to study the variations in speed of the earth's spin: F. R. Stephenson, *Historical Eclipses and Earth's Rotation* (Cambridge: Cambridge University Press, 1996). Also by Stephenson, see the chapter "Modern Uses of Ancient Astronomy," in *Astronomy Before the Telescope*, ed. C. B. F. Walker (London: British Museum Press, 1996), pp. 329–341.

Halley and the eclipse of 1715: O. Gingerich, *The Great Copernicus Chase* (Cambridge: Cambridge University Press, 1992), chap. 19.

The collective work *Roemer et la Vitesse de la Lumière* (Paris: Vrin, 1978) groups documents and testimony about the discovery of the speed of light.

XV The Shadow Line and the Shadowy Rays

The reference work for shadow in painting is M. Baxandall, *Shadows and Enlightenment* (New Haven, Conn.: Yale University Press, 1995). This is an excellent work, focusing on the idea of shadows as tools for perceiving objects' shapes.

Pliny and the lover's silhouette: *Natural History,* 35, pp. 15 and 43. V. Stoichita, in *A Short History of the Shadow* (London: Reaktion, 1997), dwells on the implications of this mythical origin of painting. The book has many beautiful pictures.

On Greek figurative art and on painting as shadow art: P. Moreno, *Pittura Greca* (Milan: Mondadori, 1987). A different interpretation of *skiagraphia* is discussed by E. Keuls in *Plato and Greek Painting* (Leyden: E. J. Brill, 1978), chap. 4.

The House of Augustus: G. Carrettoni, "La Decorazione Pittorica della Casa di Augusto sul Palatino," in *Bollettino dell'Istituto Archeologico Germanico, Sezione Romana* 90 (1983), pp. 373–419. On the use of expedients for chiaroscuro, see E. H. Gombrich, *Art and Illusion* (London: Phaidon, 1962), chap. 1.

On the removal of cast shadows in painting, see the brief but illuminating *Shadows* by E. H. Gombrich (London: National Gallery Publications, 1995).

On anchoring shadows and the way they indicate an object's height above the ground: D. Kersten, D. C. Knill, P. M. Mamassian, and I. Bülthoff, "Illusory Motion from Shadows," *Nature* 379 (1996), p. 31. The experiment by Kersten and others shows a square above a checkerboard. If the shadow is shifted only slightly, the square seems to lift up. But this seems to be a limited phenomenon—the shadow cannot get too far away. I've spent a long time atop a surveillance tower watching the shadows of seagulls fishing on the opaque surface of still water, and I've never had the impression of any immediate information gleaned from their shadows.

Hering and the shadow line: *Outlines of a Theory of Light Sense* (Cambridge: Harvard University Press, 1964), p. 8.

On the perception of shadow, the classic treatise is L. Kardos, *Ding und Schatten* (Leipzig: Barth, 1934). Kardos is extremely attentive to the way we talk about shadows—for example, he observes that we don't usually think of shadows as having a color. On shadow profile: J. M. Kennedy, *A Psychology of Picture Perception* (San Francisco: Jossey-Bass Publishers), 1974.

The contents of Leonardo's lost book on shadows are on folio 250 of the *Codex Atlanticus.* On shadowy rays, see also A. Agostini, *La Prospettiva e le Ombre nelle Opere di Leonardo da Vinci* (Pisa: Domus Galileiana, 1954), pp. 26–28, regarding a manuscript held in the Institut de France, folio 97 verso and 102 recto. On Leonardo's theory of shadows, see the excellent appendix to the Baxandall text mentioned above. On Leonardo and colored lights, see Martin Kemp, *The Science of Art: Optical Themes in Western Art from Brunelleschi to Seurat* (New Haven, Conn., and London: Yale University Press, 1990).

Talmy on shadows and "shadowons": "Fictive Motion in Language and 'Ception,' " in *Language and Space,* ed. P. Bloom et al. (Cambridge: MIT Press, 1996), pp. 211–276.

Leonardo's analogy between eye and lamp: *Codex Atlanticus,* folio 204; Agostini, op. cit., p. 33.

Piaget and Inhelder's experiments with drawing shadows: *The Child's Conception of Space* (New York: Norton, 1967), chap. 7, "The Projection of Shadows."

XVI Shadow Webs

Dürer: Albrecht Dürer's *Unterweysung der Messung,* ed. Alfred Peltzer (Munich: Süddeutsche Monatshefte, 1908). I don't discuss Filippo Brunelleschi's mirror method, which is a bit different but presents the same problems as the window method.

The birth of perspective in shadow projection: G. Bauer, "Experimental Shadow Casting and the Early History of Perspective," *Art Bulletin* 69 (1987), pp. 211–219.

Alberti: *De Statua,* ed. O. Morisani (Catania: Università di Catania, 1961), p. 47.

On the vicissitudes of Ptolemy's book: S. J. Edgerton, Jr., "Florentine Interest in Ptolemaic Cartography as Background for Renaissance Painting, Architecture, and the Discovery of America," *Journal of the Society of Architectural Historians* 33 (1974), pp. 274–292.

The dinner meeting of Toscanelli and Brunelleschi: "Filippo Brunelleschi" in Vasari's *Lives.*

The role of optics and ancient geometry in the Renaissance: K. H. Veltman and K. D. Keele, *Linear Perspective and the Visual Dimensions of Science and Art* (Munich: Deutscher Kunstverlag, 1986).

Toscanelli's role: A. Parronchi, *Studi su la Dolce Prospettiva* (Milan: Martello, 1964), and Eugenio Garin, "Ritratto di Paolo dal Pozzo Toscanelli," *Belfagor* 3 (1957), pp. 241–257.

Biagio Pelacani's role: T. DaCosta Kauffmann, "The Perspective of Shadows," in *The Mastery of Nature* (Princeton, N.J.: Princeton University Press, 1993), pp. 49–78.

A text with descriptions of many works on shadow, stretching from art to math: A. de Rosa, *Geometrie dell'Ombra* (Milan: Città Studi Edizioni, 1997).

M. Baxandall, *Shadows and Enlightenment,* shows quite convincingly that when analyzing post-Renaissance painting one should not consider only a purely geometric conception of shadow.

XVII Shadow Lessons

A reference work on Desargues: J. Dhombres and J. Sakarovitch, eds., *Desargues en Son Temps* (Paris: Blanchard, 1994).

Leibniz on perspective and shadows: *Suggestions for Establishing a General Science* (1686), in Leibniz, *Vorausedition,* bk. 6 (Münster: 1987), pp. 271–272; cited in Javier Echeverría, "Leibniz, Interprète de Desargues," in *Desargues en Son Temps,* ed. J. Dhombres and J. Sakarovitch, pp. 283–293, on p. 291.

The Discovery of Shadow

On the role played by attention in the phenomenology of shadow: M. Baxandall, *Shadows and Enlightenment* (New Haven, Conn., and London: Yale University Press, 1995), par. 39.

On the metaphysics of newborns: F. Xu and S. Carey, "Infants' Metaphysics: The Case of Numerical Identity," in *Cognitive Psychology* 30 (1996), pp. 111–153.

An extensive bibliography and documents relevant to this text can be found at the Web site www.shadowmill.com.

ACKNOWLEDGMENTS

This book was born of a conversation with Marco Vigevani: I wanted to write a short article on shadows, and he convinced me to write a whole book. For me it was the start of an extraordinary series of trips through space and time, and I wish to thank the people and the institutions who helped me in this voyage:

The team at the Centre de Recherche en Epistémologie Appliquée in Paris, particularly Jean-Pierre Dupuy, Daniel Andler, Elisabeth Pacherie, and Jérôme Dokic. At the State University of New York in Buffalo, Barry Smith, David Mark, and Len Talmy; the National Center for Geographic Analysis and Information, and the Center for Cognitive Science at SUNY. Alain Michel of the CEPERC in Aix-en-Provence; Andrew Frank and the group at the Technische Universität of Vienna. At the Warburg Institute in London (the greatest wonder of the bibliographic world): Paul Taylor, François Quiviger, and Anita Pollard. I was permitted to visit various sites and to study some hard-to-access material through the courtesy of Irene Iacopi and Gianna Musatti of the Soprintendenza Archeologica of Rome; of Father Marino Maffeo of the Specola Vaticana; of Mr. Fraiani, restorer at the Archivio Vaticano; of Father Carlo of the Basilica of Santa Maria Maggiore in Florence; of Marco Beretta of the Istituto e Museo Fiorentino di Storia della Scienza; of Annik Roger of the CNRS in Aix-en-Provence. I often found myself in unfamiliar territory where it's unwise to travel without a guide. I'm grateful for the help offered to me in those areas by Francesca Bizzarri and Fabrizio Montecchi regarding shadow theater; Maurice Bloch and Dan Sperber for the illumination about anthropology; Jonathan Raper, my mentor in geographic issues; Marco Panza for his valuable help on the history of mathematics; Ted Pedas for the journey into the heart of darkness. I was deeply influenced by conversations with Paolo Bozzi, Vittorio Girotto, and Ira Noveck on the psychology of perception and of thought. Paola Fontanini and Flaminia d'Andria followed dozens of trails across the shadowy terrain of classical philology. Heartfelt thanks to Maurizio Giri, Gloria Origgi, Walter Criscuoli, Kevin Mulligan, Dominique Fioraso, Alain de Chevigné, Paolo Legrenzi, Maria Sonino, Monique Canto, Stephan Winter, Jean Petitot, Len Talmy, Steven Davis, François Recanati, Milena Nuti, Giovanni Piana, Françoise Longy, Marisa Basile, Philippe de Rouilhan, Francesco Fagioli, Adriano Palma, Roy Sorensen, Pierluigi Ciucci, Marta Spranzi, Susan Carey, Maria Casati, Gianna Zuradelli, Pietro Corsi, Luca Bonatti, Jean-Maurice Monnoyer, Barbara Tversky, Libero Sosio, Valeria Viganò, Andrea Bonomi, Mike Martin, Ettore Cingano, and Luisa Dolza; thanks to Adriana Zangara and Davide Luglio, to Dani and Manu, to Ale and Tullio, Inma and Tito, Holly and

Matteo, Rez and Albe, and the Imbe gang. I'm grateful to the editorial department of Mondadori, particularly Giovanni Quochi, Renato Curti, and Giacomo Callo. I'm grateful to Abigail Asher for her patience with a long translation. I'd like to thank Agnes Krup, who believed in this book back when it was nothing but a list of ideas on a sheet of paper. I don't have words enough to thank Achille Varzi, my brother Marco and sister Chiara, and my parents. This book is dedicated to the near-century of Angela Farina (1907–1998), who follows me always, like a shadow.

INDEX

Sources of the Illustrations

8 Jean-Baptiste-Siméon Chardin, *Copper Urn*, Louvre. (Graphic manipulation by Walter Criscuoli)

23 Peter Schlemihl allowing his shadow to be chopped off. Engraving by George Cruikshank, 1823.

31 M, the Düsseldorf monster (Peter Kurten), with a victim, in Fritz Lang's film (1931).

34 From S. Spaggiari, L. Malaguzzi, and M. Dolci, *Everything Has a Shadow Except for Ants* (Municipality of Reggio Emilia, 1990), p. 34. Photograph: S. Sturloni and L. Vecchi.

39 Redrawn by Van de Walle, Rubenstein, and Spelke, 1998.

72 Almanac published in 1497 by K. Kachelofen, Leipzig.

84 *Astronomie Populaire* by Camille Flammarion, p. 225 of the 1908 edition.

94 Peary pretending to pose at the North Pole. From National Geographic Society.

104 Giovanni Battista Piranesi, "View of the Renowned Vatican Basilica with its Wide Portico and Adjacent Piazza," 1773, in *Vedute di Roma*, vol. I, pl. 3, etching.

118 Photograph shot from *Apollo 12*. (NASA)

119 From Camille Flammarion's *Astronomie Populaire.*

131 The phases of Venus seen by Kepler, *Epitome of Copernican Astronomy*, pt. III.

133 Mercury, seen by Gassendi on November 8, 1631. From *Mercurius in Sole Visus.*

139 Saturn, viewed by the *Voyager* probe on October 30, 1980. Photograph: NASA, AC80-7023.

151 From *Journal des Scavans*, December 7, 1676.

158 Room of the Masks in the House of Augustus on the Palatine Hill. Soprintendenza Archeologica di Roma.

160 Filippo Lippi, *Nativity* (detail), Duomo of Spoleto.

161 Luca Signorelli, *Dante's Shadow*, San Brizio Chapel, Orvieto. From *Temi Danteschi a Orvieto* (Milan: Arti Grafiche Ricordi, 1965), p. 51.

163 Masaccio, *Saint Peter Healing with His Shadow*, Brancacci Chapel, Santa Maria del Carmine, Florence.

164 Masaccio, *The Tribute Money*, Brancacci Chapel, Santa Maria del Carmine, Florence.

A Note About the Author

ROBERTO CASATI was born in Milan in 1961. A research director at the CNRS, he lives in Paris and works at the Institut Nicod, a laboratory of the Ecole des Hauters Etudes en Sciences Sociales and of the Ecole Normale. He studies the cognition of strange things—images, colors, sounds, places, holes—and shadows. With Achille Varzi he is the author of *Holes and Other Superficialities* and *Parts and Places.*

A Note About the Translator

ABIGAIL ASHER is an editor of art history and other nonfiction books. She has lived and worked in publishing in Italy; she now lives and works in New York City. This is her first book-length translation.

A Note on the Type

This book was set in Minion, a typeface produced by the Adobe Corporation specifically for the Macintosh personal computer, and released in 1990. Designed by Robert Slimbach, Minion combines the classic characteristics of old-style faces with the full complement of weights required for modern typesetting.

Composed by Creative Graphics, Allentown, Pennsylvania
Printed and bound by Berryville Graphics, Berryville, Virginia
Designed by Anthea Lingeman